The Spirit of Beautiful Trees

R. A. RAJU
M. Sc. (Ag), Ph.D., FFSRDA
Acharya N.G. Ranga Agricultural University
Agricultural Research Station
Maruteru – 534 122 (India)

Agrobios (India)

Published by:

Agrobios (India)

Agro House, Behind Nasrani Cinema
Chopasani Road, Jodhpur 342 002
Phone: 91-0291-642319, Fax: 643993
E. mail: agrobios@vsnl.com

Agrobios (India)

ISBN ...-2

9 788177 540888

Rs. 1295.00 / US$ .00

Published by:
AGROBIOS (INDIA)

Pri... 00

Suggested Citation: Raju, R. A. (2002) The Spirit of Beautiful Trees. Agrobios (India), Jodhpur – 342 002, 221p

Published by: Updesh Purohit for Agrobios (India), Jodhpur
Laser typeset at: Commerce House, Jodhpur
Paintings and Line Illustrations: Parvati, N. S.
Cover Design by: Reena
Printed at: Anmol Prints, Jodhpur

Dedication to,

Lord Sri Heramba

Who is affectionate son of
Soundaryalahari and
Neelakanteswara
The elder brother of Kartikeya

Who had elephant head and
Naga as sacred thread

The Lord of all obstacles and
Embodiment of wealth and wisdom

"The landscapes beautified by a variety of flowering trees,

Cortal red flowers of Erythrinas, Golden showers of Indian Laburnum,

Vermillion blooms of palas, deep mouve floral clusters of Jacarandas,

Blazing blooms of Frangipani, honey scented white neem inflorescences,

Large opulent purplish-red flowers of red-silk cotton,

Sacred peepal trees near shrines,

Always remembers the fascinating Indian trees,

Glistening blue breast of peafoul,

Brilliant blue wings of king fishes,

Melodious singing cuckoos,

Snow-white winged ducks and feathers of painted storcks of marshes,

Display the marvel of miracle nature in India,

It is feast of the eyes and an unforgetful sight to every foreign traveller."

Foreword

Dr. M. S. Swaminathan
Ex. Director General, ICAR, New Delhi
Ex. Director General, IRRI, Manila, Philippines
Ex. Vice Chairman, Planning Commission of India
Chairman, M.S.S. Research Foundation
Third Cross Street
Taramani Institutional Area
Chennai – 600 113, (India)

Trees are the most precious gifts of nature to human kind. The multiple roles they play in helping to improve the quality of human life have been described by many poets and saints. In the context of global warming, there is increasing interest in trees from the point of view of carbon sequestration. Thus, the present publication on *"The Spirit of Beautiful Trees"* by Dr. R. A. Raju is a timely one.

We need to creat greater interest in trees among our young students. Nearly 25 years ago, I wrote to all our Universities and Forest Departments asking them to identify the oldest tree in the area. Most of them wrote back describing a banyan tree in their neighbourhood. I however wanted to locate the oldest living tree of India. After a great deal of search, a *Morus Serrata* tree near *Joshimath* was identified as being over 1300 year old. The age of this tree was related to the visit of *Adi Sankara* to that area, who conducted prayers under that tree. Thus, trees in our country are not only beautiful but also serve as spiritual flames. I hope this book will help to spread a love of trees in every one, both young and old, male and female and will contribute to the greening of India

M. S. Swaminathan

Front line

Trees are natural assets of immense value to any country. They not only preserve the physical features of the earth, stop soil erosion, mitigate floods but also make the streams flow perennially and help in sustaining river flows. Trees also make the climate equable, cool and encourage the precipitation from clouds. They provide shelter to wild life and support the biological chain. Tree cover helps in maintenance of ecological balance of nature. Besides, Trees meet the needs of timber, fuel, medicines and other commercial products which are indispensable requirements of human beings. Trees are natural heritage of the country and they represent the remarkable biological diversity and genetic resources of the earth.

The clean environment and beautiful surroundings of nature elevates human emotions and fills the hearts with joy. Beautiful trees are valuable asset to any country because of their multiple utility. They symbolizes beauty of nature and make the inert atmosphere into live environment. The gracy flowering trees display their colourful blooms in various times of the year and indicates the advent of seasons and their significance. Planting of flowering and fruit trees has been popular in India since antiquity.

Beauty not lies in massive buildings and huge structures. Surrounding vegetation removes drab environment and adds graciness to buildings. Some of the majestic trees like peepal, neem and mahua are very hospitable. Apart from humans, they attract a large number of visiting birds and other interesting creatures and gives boarding for them. The reckless distruction of trees continuing day after day as the burgeoning population occupying more and more land. Several species of trees and wild life are under threat of extinction.

To sustain the tree resources and wealth it is necessary to develop programme of the education. Keeping these aspects, the author of this book presented the matter coherently on various aspects of flowering trees present in India for the benefit of town planners, architects, builders, administrators and students engaged in agriculture. I sincerely hope that the present book written by Dr. R.A. Raju should be useful to nature lovers, urban planners and layman alike.

I congratulate the author for its efforts in helping to conserve the aesthatic value of our rich tree wealth.

Panjab Singh
Director
Indian Agricultural Research Institute,
Pusa, New Delhi – 110 012
Ex. Vice-Chancellor,
J.N. Agricultural University
Jabalpur

Preface

A rich and healthy nation requires not only natural resources, good climate, disciplined people but also beautiful environment surrounded by green trees. The trees having brilliant coloured flowers play a pivotle role in making the surroundings Joyful and Spiritual. Such a beautiful environment refine the minds of people and activate the moods. The beautiness of a city or town not determined by its vast area but by its bioaesthetic planning along with majestic trees. Ornamental flowering trees along the roads of cities, houses and public places acts as bird sancturies and helps in preservation of graceful avion fauna. The colourful birds and flowers remove the drab environment of concrete jungles and rise the spirit of city people.

Fortunately, India has a vast wealth of bright beautiful flowering trees. Trees like blue flowered Jacarandas, golden yellow amaltas, orange-scarlet gold mohars, dazzling scarlet colvillias, blood red eirthrinas and pink largerstroemias are unforgetful if they are seen in full blooms. Proper planning is essential to plant the trees suitable to the environment. Besides planning, continuous efforts are needed for growing these trees in a scientific way. The beauty of Landscape gardens of Japan, maple groves of Canada, the vineyards of Rhine, tulip fields of Holland, saffron terraces of Kashmir valley, hedge rows of England, grand avenues of Srinagar Chenars, poplar-lined roads of France are not creation of nature. These excellent things are the result of planning and hardwork of several generations.

Ancient people revered the trees and believe that spirits dwell in them. Some tribals organise festivals when certain trees like *Sal* blooms. At the time of Buddha's birth, his mother Mayadevi stretched out her hand to hold a branch of sal tree during her delivery of child. Sal trees are also said to have rendered homage to Buddha at his *nirvana* by offering their flowers fall on him out of a season. The sal branches bended on him to provide shade from the hot sun. There is a tradition of worship of these trees by buddhists and Hindus.

The reckless destruction of trees continuing day after day as the burgeoning population occupying the limited land resources. Several species of trees and wildlife are under threat of extinction to sustain tree resources and wealth, it is necessary to develop programme of tree culture. People should be educated and inculcate love on tree planting.

I was inspired by the works of Late Dr. M.S. Randhawa, and H. Santapau who loved the tree wealth of India. The book presented on various aspects of charming trees for the benefit of town planners, architects, builders, teachers and taught of agriculture and horticulture and extention personnel who engaged in establishing green belts. The author express his thanks whole heartedly to Mrs N. Sita Parvati for her excellent illustrations and paintings. My deep appreciation go to Dr. I.V. Subba Rao, Vice-Chancellor, A N G R A U, Hyderabed for the encouragement and helpful comments on the project. I hope that the present book should be useful to all sects of people. Suggestions are welcome from the readers for revising the future edition.

R.A. RAJU

Contents at a Glance

Chapter

1

Introduction

"GOD HAS MADE EVERYTHING OF NATURE BEAUTIFUL AND ROMANTIC IN ITS TIME"

The flowering trees are symbolic of beauty, love and tranquility. They beautify the surroundings and elevate the human emotions. The nature conveys its happy message and subtle love to her children in various seasons through flowering trees. Planting of flowering and fruit trees has been popular in India since ancient times. According to *Valmiki Ramayana*, the beautiful '*Ashokavana*' was existed in which *sita* was in captivity. The shade giving Ashoka trees were predominant in this royal garden. Descriptions about layout of parks, fruit gardens and artificial lakes in the capital city of *Indraprastha* is made in the *Sabhaparva* Chapter of the Mahabharata. In Ramayana, mention has been made about several flowering trees, such as, *Mechalia Champaka* (Champak) *Terminalia arjuna* (Arjun), *Saraca indica* (Ashoka), *Mesua ferrea* (Iron wood), *Calophyllum inophyllum* (Nagachampaka), *Butea monosperma* (Palas), *Cassia fistula* (Inidian laburnum) and *Ficus religiosa* (Peepal) The intimate association of lovers *Radha* and *Sri Vamsi* with *Kadamba* (*Anthocephalus indicus*) trees present on the bank of Yamuna river is well known to everybody in Mahabhagavata.

1.1 Historical background

According to Buddhistic literature, Lord Goutama Buddha was born under *Palas* (*Butea monosperma*) tree in Lumbini and attained *Nirvana* under the Pipal tree called *Maha Bodhi*. The poet Aswaghosa (100 AD) vividly written about *Nandanavana* in which Siddhartha seen the beautiful flowering trees and lotuses in ponds. The kings who followed Buddhism at that time built the monastries and *Stupas* and beautiful gardens were laid out around them. The emperor Ashoka the great (233 BC) planted the shady avenue trees "*Margesu-Vriksha*" for the benefit of long distance travellors. The Sanskrit play writer Sudraka (100 BC) has also described about flowering trees and garden management in his play '*Mrichhakatikam*'. It is said that majestic gardens were present at the educational institutes of Nalanda and Texila.

1.2 Literary evidence

Kalidasa, the great Sanskrit poet (57 BC) described some of the romantic trees and shrubs like Vakula (*Mimusops elengi*), Parijata (*Nyctanthes arbor-tristis*), Madhavi lata (*Hiptage madablota*), and Kavidhara (*Bauhinia variegata*) in his play *Shakuntala*. Interesting accounts of different kinds of tree gardens like *Vrikshavatika, Udyan, Nandanvan* and *pramadodyan* were mentioned by Vatsayana (300 –400 AD) who was *written Kamasutra* (The principles of sex life). The art of tree planting and ornamental gardening has been described by Sarangadhara, the prince of middle ages, in his works *Upavana Vinoda* and *Sarangadhara paddati*.

The ornamental trees and flowers also found in architecture of Mathura during Kanishka period at Bharhut, Sanchi and Ajanta. According to ancient sanskrit scripts the indigenous flowering trees in India are Tulip (*Thespesia populnea*), Scarlet cordia (*Cordia sebestena*), Indian coral (*Erythrina blakei*), Way leafed Tecomellia (*Tecomella undulata*), Chalta (*Dillenia indica*), *Crataeva roxburghii* and Dhak (*Butea frondosa*) which were grown in the gardens of ancient kings.

1.3 Renaissance of ornamental gardening

The native rulers did not showed much interest in developing ornamental gardening in public places. Mughal emperors have a great interest in developing beautiful gardens and introduced several exotic flowering plants. The first Mughal ruler Babar brought several ornamental plants from Persia and central Asia. These are mentioned in the autobiographics of that period. Mughals introduced *Chenar* trees of Kashmir, roses, iris, tulip, lilies, daffodels, carnation, narcissus *etc* into India from Central Asia. Emperor Jehangir and Nurjahan encouraged the gardens because of their deep love towards roses and other flowering plants.

1.4 Archaeological evidence of tree groves

Man was evolved at the advent of pleistocane age, about a million years ago. The earth was covered with thick stands of large trees except in Rajputana region and parts of *Panchala* which lay buried under a marsh, the remnant of the receding Tethys Sea. Rajputana marshes existed till 4000 B.C. When Mohenjodaro Culture flourished in the out-skirts of Lothal in Gujarat. Archaeological evidence indicates that stout reeds grew in these swamps which are used to cover the dead bodies in the Chalcolithic age. These are evidences that rhinoceroses are roamed in the neighbouring forests. Fossilized crocodilies are also appeared in these places, indicating the existence of water ponds and streams. At this time, the thick stands of trees might have been cutted by bronze axes and destroyed the forests. The great Civilizations disappeared, possibly due to destruction of vegetation cover which results in eroded fertile soils.

1.5 Worshipping of Nature by Aryans

Aryans entered into India after 2000 BC and started using iron ploughs, Javelins and axes. They are primarily pastoral people and cleared the big trees and forests for the construction of houses, godowns and shelters for the domesticated animals. However, Aryans are worshippers of Nature and build their abodes, *Ashrams* and *tapovatikas* in sylvan surroundings.

They composed *Vedas*, *Upanishads* and *Aranyakas* which sing the glory of Universal Lord. These works laydown precepts of conduct for human society to live righteously and peacefully.

1.6 Ancient Hindu Phylosophy

In the great epic *Ramayana*, descriptions were made about dense forests of *Dandakaranya*, *Naimisaranya*, *Chitrakoot* and *Panchvati* which are teemed with wild animals. Even in Mahabharata one can come across the tree groves like *Naimisaranya* and *Khandavavana*. To arrest the reckless cuttings of trees and burning of forests, ancient people declared that destruction of tree is an act of sin. According to *Agnipurana* the planting of tree is an act of piety.

"Das Koop Samavapi, Dasvapi Samahad
Dashad Samaputro, Dasputra Samataru"

"Digging of ten wells are equal to one pond
Ten ponds are equal to made one lake
Ten lakes are equal to birth of one son
Ten sons are equivalent to plant one tree"

Several Hindu philosophical treatises banned the extinction of trees. They worshipped the trees by knowing their role in maintaining microclimate, water cycle and condensation of clouds.

1.7 Stories of foreign pilgrims

Chinese pilgrims like *Huan-Tchang* and *Pahiyan* visited India during middle ages to study the Buddist philosophy and written their experiences in books. These chronicles are the base of knowledge about the fate of trees, forests and wild life at that time. The great Greek warrior, *Alexander* observed the very high dense forests which are impenetrable to the human along the Indus river. The emperor *Mourya Chandra Gupta* planted new species of trees in open lands and arrested the Vandalism of cutting trees recklessly. According to *Chanakya's*

(Kautilya) *Arthasastra*, a special officer was appointed by the king to protect the forests and wildlife.

Later inroads had been made into forests for the transport of forest products and to expand the agriculture due to population pressure. Emperor *Ashoka* felt the need of heavy roads for the transport and business and arranged for planting shady trees along the highways. The great monarch also encouraged the planting and growing of exotic medicinal trees on either side of roads and camping grounds.

1.8 Interest of tree planting by invaders

The Afghan ruler *Shersha Suri* planted the magnificent trees along the road of Delhi-Patna. Although Mughal emperors are not interested in growing forest trees but planned exquisite gardens in Delhi and Kashmir. The *Padusha* Jahangir brought Chinar trees from Afghanistan and introduced it in Kashmir valley. These Chinars became famous and become the synonymous with Kashmir people. The historical record *Ain-e-Akbari* mentioned about growing of hunting grounds for the hunting of wild animals like tiger and lion. It also mentioned that hundreds of elephants are roamed in the forests of Mhow, Madya Pradesh. Kings of *Maratas* and *Gonds* planted fruit trees like mangoes, Tamarind and other trees along their marching routes of armies.

1.9 Biggest trees of the Universe

Some of the trees grown in wild habitats attains to a height of more than 100 metres. The biggest tress in the world are present in california and Southern oregon in U.S.A. The coastal Redwoods (*Sequoia sempervirens* End) are largest untouched stands flourished in Humboldt and Del Norte counties of Northern california. The biggest big tree is called *General Sherman* which is found in U.S.A. If this tree is slice off at its base, one can found enough room in its stump for a small house. It requires to join 25 tall persons hands to encircle the tree. This will give the idea of its stem girth. The tree attained to a height of 83m, which is an equivalent height of a 25 storey skyscraper. If the stem is cut-off into planks, it may fill about 200 railway coaches. Its weight is estimated to be about 6000 tonnes. It is surprising to note that its age is 4000 years old and people believe that the tree is odest living things on the earth. It is estimated that it could live for atleast more than 2000 years.

1.10 Monarch of Foreign Lands

Other monarchs of tall trees are redwood creek grove of Humboldt county (111m), Rockefeller tree of Humboldt (107m), Founders tree (106m) and redwood creek Grove (112m). The species like Douglas fir (*Pseudotsuga taxifolia*) registered 98m at Ryderwood in Washington and *Eucalyptus regnans* (96m) in Victoria, Australia. They are eager to touch the clouds.

In India, some protected largest trees are *Dipecerocarps gracilis* (44m) in Andamans, *Eucalyptus globulus* (37m) in Kodaikanal, *Pinus roxburghii* (66m) in U.P, *Pterocarpus dalbergiodes* (52m) in Andamans, *Shorea robusta* (42m) in U.P and *Tectona grandis* (26m) in Karnataka. Although these trees are small when compared with california redwoods but many become future gaints on the earth.

1.11 Tree with bold trunks

Few trees; over hundreds of years and increase their girth amazingly due to continuous growth process. Some of the fatty trunk trees with maximum girths in India are Blue Gum (5.4m) in Niligiris, Kashmir, Chinar (21.5) of Bijbehara, Garhwal Chir pine (4.2m) of U.P, Tone Deodar (8.1m) of U.P, Malda mango (7m) of West Bengal, North Andaman Paduak (6.1m) of Andaman islands, Haldwani Sal of U.P. (2.7m), Garhwal spruce (7.4m) of Uttaranchal, South Chanda Teak (4.0m) of Maharashtra and Bilaspur Laurel (5.7m) of Madya Pradesh.

The Spirit of Beautiful Trees **Plate 1**

Fig. 1:
Who on earth has thought of that there is no marvel in nature,
Has the divine king enjoy more than this spirit of amazing trees?
No human feel exited by seeing the glorious display
of Blue oak lands of California

Fig. 2:
Wild almonds while displaying their full blooms.

Fig. 3:
Woodman, spare that tree ! Touch not a single bough !
In youth it sheltered me ! And I will protect it now
- *George Morris*

Fig. 4: Wild Cherries with dazzling scarlet fruits

"Loveiest of trees, the cherry.
Filled with delicious sweet juice.
Hung with a handful of ripened
fruit along with delicate bongh"

Fig. 5: Chirpines on snow covered mountain slops.

Fig. 6:
Lush green tree create shade, protect the atmosphere from air and noise pollution, provide wildlife habitat, as well as make the surroundings highly attractive and feast the eges of viewers.

Fig.7:
Display of bright lemon yellow flowers against the dark green foliage

Fig. 8:
Bhendi
(Thespesia populnea)

L - Dark green
Fl - Yellow
Fr - Brown

Fig. 9:
West indian mountain rose
(*Brownea coccinea*)

L - Dark green
Fl - Scarlet red
Fr -

Fig. 10:
Barna
(*Crataeva religiosa*)

L - Green
Fl - Plae yellow / Reddish Yellow
Fr - Scarlet red

Fig. 11:
Moulmein Rosewood
(*Milletia ovalifolia*)

L - Green
Fl - Mauve / Lilac
Fr - Brown

Fig. 12:
Arjuna (*Terminalia arpine*)

L - Green
Fl - Pale yellow
Fr - Tan
Grey bark

Fig. 13:
Pilkhan (*Ficus religiosa*)
F - Green (timted copper)
Fl -
Fr - White
Wood - grey

Fig. 14:
Buddleja pamiculata
The ripened orange red leaves and dense, snuffy inflorescence transform the surroundings romantic

Fig. 15:
Variegated Bauhinia (*Bauhinia variegata*)

1.12 Trees of *Tretayuga*

Most of the trees are short lived but few trees live for hundreds of years and grows continuously like Deodar. Banyan tree can be termed as *Akshaya Vat* (indestructible) because it throughs out its aerial roots produced from branches into soil and spread continuously. The Sibpur great Banyan tree is more than 200 years old and stands as main attraction in Indian Botanical Garden in Calcutta. Bristlecone pines (*Pinus aristate*) found in white Mountains of Eastern california are found to be more than 4000 years old. In order to preserve the oldest trees the U.S. Forest service has set aside 8000 hectares of public lands in the White mountains.

1.13 Ancient Bristlecone Pine Area

The longevity of some of the trees are surprizing and unbelievable. They are yew tree (more than 3000 years) of England, Tehri-Garhwal Deodar (710 years) of Uttaranchal, Ulandi Teak tree (over 500 years) of Tamilnadu, Ulandi Shisham (over 600 year) of Tamilnadu, Chirpine of Chakrata (350 years) hills, and Srinagar Chinar (300 years) of Kashmir. The longlife of these oldest living things on earth is a mystery. These may be attributed due to more pith in its trunks, presence of Chemicals in pith which are resistant to insects and rot, and very slow growth rate (2.5cm radius in 100 years).

1.14 Sacred thoughts about Venerable trees

People who leads to pious life and work untiredly for the society are considered as sacred men. Similarly some of the trees like Peepal, Bael and neem are considered highly sacred and worship them. If one goes to rural areas of South India and Indo-Gangetic plains, there are scenes like pouring water on the trunk of peepal in the early morning. In interior south, every village has one peepal near the canal or river and people worship it after completion of their bath in the canal. When the new moon falls on a Monday, the peepal is worshipped by all women. They pore pot full of water at the bottom, lit the oil thread and put the copper coins and sweetmeats on its thick roots and pray to grant children and wealth. They believe that it is highly dangerous to lie, cheat and stealth beneath the shade of peepal because it is a dwelling place of supreme God. Hindus believe that the peepal roots represents *Brahma*, the God of Creation, its trunk represents *Mahavishnu*, the God responsible for the growth and protection of all living beings, the branches represents *Mahashiva*, the destroyer of beings at the end. Hence the tree is considered as highly sacred. According to *Vishnu Purana*, the whole universe contained in *Brahma* like the magnificant, wide spreading peepal tree embeded in a small tiny seed. *Vishnu Purana* described the *Aswatha* Peepal as

"Moolato Brahma rupava, madyato Vishnu roopine
Agrata Sivaroopaya, Vruksha rajayate Namaha"

"The king of the trees, having Brahma,
Vishnu and Shiva in its roots,
trunk and branches respectively
I convey my respects."

1.15 Enlightment of Siddhardha under *Bodhi*

Ancient sages chose to sit and meditate under the peepal due to its sacredness. *Gautama Buddha*, the founder of Buddism meditated under the peepal tree present in Gaya and got enlightenment on the day of *Buddha Purhnima*. The same tree was called as *Bodhi* which means tree of wisdom. The sapling of tree was taken from Gaya to Srilanka in 288 BC by Mahendra and Sanghamitra, the Children of Great Ashoka. Even now, the tree is alive and carrying its growth. The age of the tree must be 2288 years old by now. The bark and dry branches used in sacred fires *(Yajna)* during festive days.

To the Hindus, the bael tree is also highly sacred tree. The tree is known to inhabitants of *Bharatkhanda* since vedic period (800 BC). Several references have been made in *Yajur Veda* and other sanskrit scriptures. According to Hindu mythology. Lord *Kailashnath* takes rest beneath the

Bilva tree. The compound leaf consists of three leaflets which represents the forms of *Trimurthi*. According to sanskrit *Puranas* the Beal was described as

> ***"Thridalam, Thrigunakaram, Thrinethram cha Triyayudham,***
> ***Thrijanma Papasamharam, Yekabilvam Samarpanam"***

which means:

"Three leaved compound leaf represents three character, three eyes, and three weapons of war, offering of one bael leaf burns the sins of three past births."

One can imagine the feelings of ancients about the sacredness of trees. They planted the sacred trees to earn the blessings of heavenly Gods and future generations as well.

1.16 Touch of beauty with gracy trees

The development of any country mainly depends upon balanced growth of both agricultural and industrial sectors. Wealthy nations of the globe already reached the goal and accumulated vast wealth in the form of material goods. Mere accumulation of material wealth did not give happiness to the people. The graceful and gallant trees, blazed colour of flowers, amazing wild animals and songs of little birds makes the environment joyful and refine the hearts of its people. Nature is a master-artist who paints various colourful pigments to the flora in different seasons. The nature lover can observe richly coloured surroundings in the country side. **The Orange-scarlet flowers of *Gold Mohurs*, hanging inflorescence of Indian Laburnum, blue Jacarandas, Pink flowers of lager-stroemias, brilliant red erythrinas and blossoms of scarlet colvillbeas create an eye-feast picture and please the viewers.**

Several poets like wordsworth and Kalidasa inculcated the love of nature through their writings. At this computer age, people began to admire the beauty of flora and fauna than the yester years. The present generation wants to enjoy life in pollution-free, dust-free, slum-free, quite, noise-free environment and green surroundings.

1.17 Restaurant of Colourful birds

Trees with multiple branches attracts several birds and small animals. Sweety singing *Koels* (Cuckoos) usually visit the mango trees comes to full bloom. Armies of parrots sit on the mango branches when it bears fruits. It is an wonderful picture when flowering trees coupled with Colourful birds. If any one visits the crimson silk-cotton tree (*Semal*) either in the jungle or in country side during February, the birds club can be seen there. The bright crimson flowers shaped like shuttlecock display the brilliance of beauty. During the start of the year, the tree become completely bare like tribal women who takes bath in the river. The leafless tree completely covers with bright mass of flowers within one month. At the end of winter, the beauty of tree-lined avenues of *semal* can be worth seeing.

1.18 Music Club of Mynas

A large number of colourful birds visit the tree due to colour and profusion of flowers. The lively chatter of mynas, rosy pastors, bass of crows and twittering golden sparrows and honey suckers make the tree like musical club. The mynas come to sip the nectar found in the big, widened, crimson flowers. Barbets, bulbuls, koels in groups visit the flowers for breaking their night fast. Some king crows frequent the tree for taking the feast of lunch. Some blue coloured sparrows come in search of hundreds of drowned insects which settled in the bottom of flower cups. Wood peckers visit the tree to hollow the soft stem for making nest. Besides the birds squirrels run on branches and chattering noisily here and there. Thus there is a rush of little birds on the tree from dawn to dusk.

The rosy pastors are also significant visitors of the trees but not singly. Flocks of birds settle on the branches and start the sipping of honey. The flowers and rosy pastors are similar in colour and one can not identify them easily. The voices of pastors are so loud that the people think that some riot is going on. This grand symphony of nature is highly exciting.

1.19 The mesmerism of mango groves

The Mango tree enjoys a prominent position among the Indian Trees. The taste and uses of mango is unique and known as ***"King of tropical fruits"***. Even in wild state, the tree is so generous and gives juicy fruits and cool shade to the travellers during scorching summers. Ancient sanskrit literature described the fruit as ***"the nectar of Gods"***. This stateliest tree has been found in India since vedic period. According Hindu mythology, Lord *Mahashiva* created the mango fruit with his divine power for presenting to the daughter of Himavanta. The tree is regarded as highly sacred and auspicious, Mango juice is poured on *Shiva Linga* (Abhishek) and worshipped on mondays in *Saiva* temples of South India. The devotees believe that Maheswara enjoys the bath of mango juice and fulfill their wishes.

1.20 Mouth watering mango pickles

The Greek conquerror Alexander the great, saw the mango bearing trees and appreciated their taste when he invaded the North Western part of India. Persian Voyagers delighted with the taste of juicy fruit and introduced into African coasts. No fruit can substitute the Kitchen requirements of mango in South Indian houses. It is so famous as vegetables and fruit as well. The green mangoes cooked with pulses and vegetables like brinjal and eaten during summers. The unripened mangoes are pickled in various forms by mixing with gingelly oil. There is an hectic activity of preparation of different types of mango pickles in every home of South India during hot summer. These pickles are preserved more than one year, used daily when taking meals. The juice is also dried on palm mats and flakes are prepared for future use.

1.21 Picnics in *Kartika* of *Sarat rutu*

Immediately after the cessation of rains, *Sarat rutu* will enter and the climatic Changes occur from cloudy to brilliant skies. In peninsular India, groves of giant mango trees can be seen in every village about three decades back. The entire village folk gather in these mango groves during *Kartika masa* (November) and cook various vegetable dishes along with rice and offered to Lord Shiva. At that time the entire grove will be like a mini village. The mouth watering flavours diffused into the air from various cooking dishes. Children will be very busy with the collection of firewood from dry branches for firing the hearths. All the people sits down under the mango tree and eat their feast, following it with some petty games upto the evening. The groves are busy with folk feasts and songs in the entire month.

If one goes to South India, he can see the mango groves on the side of highways like oasis in the desert. Beneath the trees either open well or borewell may present on one side. The farmers and farm labour working in Sugarcane and elephant yam plantations sit beneath the trees and eat their rice in the hot days. These groves are very, very old and they are not able to say in which past generation they have been planted. The town *Hazaribagh* of Bihar is full of mango groves in ancient days. The name *'Hazaribagh'* means ***"Garden of thousand trees"***, Modern inhabitants of the town cutted the trees for the expansion of the town.

1.22 Feast for parrots

Mango groves are raided by parties of parrots during the fruiting period. These swarms sit on the top branches, bite deep into the fleshy green fruit. They damage nearly half of the fruits before come to ripe. When the farmer shout with loud voice, the emerald green parrots circle round the grove with long screams and again settle on the near-by garden. When the egg laying season begins, the grove is again busy with parrots, blue jays and little green coppersmiths, even though fruits are absent. They make the holes on the trunks of old mango trees and used as nesting place. Parrots gnaw the fruits, eat little, waste enormous quantity by cutting them. Village children run into the grove to take away the parrot bitten fallen fruits.

1.23 Melodious songs of Koels

When the spring season starts, the *koels* are busy with singing from dawn to dusk. The bird is very famous for its sweet calls in the morning hours and several poets discribed these songs in their poetry. People compare the best lady singer with *Koel*. Poets attribute this melodious voice of *koel* is mainly owing to consumption of tender buds and flowers of Mango during spring season.

1.24 Blaze of bright flowers

The fresh beauty of nature freshen the minds of not only sensitive artists but also common man. The fresh breezes from neem groves, glamour of colvilleas, dazzling amaltas, glory of erythrinas, splendour of kachnars and blaze of gold mohurs during the blooming season highten the imagination of any human being. He convey his heartfelt thanks to the Almighty who created this magical and colourful universe. While planning for avenues with flowering trees one should consult with artist for suitable colour combinations. Most of the flowering trees looks elegant when planted in pure avenues. However combinations of various colours also synergise the effectiveness when planted alternatily on the side of city roads. Some of coloured combinations that harmonize the attractiveness in various seasons are given hereunder:

1. Summer flowering system

a)	Indian Laburnum (*Cassia fistula*) bright yellow	Gold Mohar (*Poinciana regia*) Scarlet orange	Indian Laburnum (*Cessia fistula*) bright yellow.
b)	*Cassia nodosa* (Pink)	Cassia marginata (Terracotta red)	*Cassia nodosa* (Pink)

2. Spring flowering system

a)	Scarlet erythrina (*Erythrina indica*) (Scarlet red)	Fountain tree (*Spathodea nilotica*) (Orange Crimson)	Scarlet erythrina (*Erythrina indica*) (Scarlet red)
b)	*Grevillea robusta* Yellow	*Jacaranda mimosaefolia* Blue	*Grevillea robusta* Yellow
c)	*Banhinia Krugii* (Light pink)	*B.Krugii* (Light magenta) (White)	*Bauhinia Variegata* *B.Variegata* (Purple mauve)

3. Autumn flowering system

a)	Colvilles glory (*Colvillea race mosa*)	Rusty shield bearer (*Peltophorum ferrugineum*)	Colvilles glory (*C.racemosa*)

4. Rainy flowering system

a)	Crepe flower (*Lagerstroemia thorelli*) White-manve	Queens flower (*L. flos-reginae*) maure purple	Crepe flower (*L. thorelli*) White maure
b)	White Frangipani (*Plumeria alba*) Snow White	Temple tree (*Plumeria acutifolia*) White with Yellow center	White Frangipani (*P.alba*) Snow white
c)	Rain tree (*Enterolobium saman*) Pale-Pink flower panicles	Rusty shield bearer (*Peltophorum inerme*) Bright golden yellow	Rain tree (*E. saman*) Pale-Pink
d)	Kadamba (*Anthocephalus Cadamba*) golden balls of flowers	Variegated Banhinia (*Bauhinia variegata*) Purple manve	Kadamba (*A. cadamba*) golden yellow

1.25 Bioaesthetic planning of Cities and Towns

Beauty seldom lies in concrete building and sky scrappers. The rich flora and fauna of the nature is the real beauty that will give soulfull happiness to the inhabitants. Shiny green trees and colourful flowers play a major part in making the environment beautiful. Planting the trees and planning the modern cities is like a music, a most sensitive fine art.

The planner should have good knowledge of ecology, morphology and physiology of trees besides architecture and sculpture of buildings. An aesthate have a deep love for nature and identify the colourful trees suited to the environment and locality. Architects of cities should have aesthatic sense to balance the rhythm and accent of flowering trees with buildings and road. Bioaesthatic planning of a city is like decoration of new bride in beauty parlour. The beautician should know the principles of body anatomy and colour and suitable matching of coloured sarees and ornaments. Although, a beautiful lady has well developed and shaped body but less attractive when she wear improper dressing. A young women expresses her glamour when the body organs are properly dressed with colourful sarees and golden ornaments. Similarly unplanned cities and buildings planned by a unwise architect creates a sense of irritation to the viewer. Landscape design is a three dimensional which involves shape of the site, suitable trees and colour of background sky. Trees like Bhendi, Magnolia, West Indian mountain rose, Barna etc (Fig 1.8, 1.9, 1.10, 1.11 not only gives shade but also rise the spirit of humans with dazzling flowers.

1.26 Changed psychologies of Citizens

Ancient people of about few centuries back are afraid of mighty mountains, dense forests, dark tree groves, chain of hills and vast lakes and seas due to some misbeliefs and taboos prevailed at that time. The country man looked these things with a feeling of fear and horror. Brutal foreign invasions and their destructive activities might have created such type of feary feeling and printed in their minds permanently. They constructed their houses in narrow lanes where the light and air cannot penetrate. These illventilated and unplanned houses have crooked lanes, narrow roads, zig-zag pathways creates crazy feeling to the dwellers. Such type of houses with no aeration and surrounding green may be suitable in the middle ages when people find no security against the invadors and thieves. Congested, unventilated houses with narrow lanes can be seen in the old cities of varanasi, Hyderabad, Lucknow, Patna,

Amritsar, Calcutta, Allahabad and kanpur. A bird's eye view reveal that these houses are like crazy puzzles.

Fig. 1.1: Magnolia (*Magnolia stellata*), L – green, Fl - White

1.27 Beauty Conscionsness of the Society

After the independence, the literate percentage of people increased significantly and the society changed its psychology. People who travelled the Europe and American continents aware of orderliness of cities and became beauty conscious.

"One can not forget the romantic
lands of Netherlands,
Poplar-lined roads of France,
Vineyards of Rhine,
Saffron terraces of Kashmir valley,
Hedgerows of great Britain,
Maple groves of Canada."

These are not nature's gift but the hardwork of human to create a semblance of order.

1.28 Gardens of Indus Valley Civilization

The earliest civilization that flouerished in India on the bank of river Indus between 3000-1500 BC is known as the *Indus valley civilization*. The discovery of this civilization was the result of excavations of the archaeological department in 1920-22 at Mohanjodaro and Harappa in the Larkana and Montgomery districts of Pakistan. It is believed that the Indus valley civilization had achieved almost accomplishment in human existence. There were magnifiwent towns which were well planned and well-built. The houses were made of bricks and had two or more rooms for storage. The people led a luxurious life. They built the gardens and irrigation and drainage canals. They ware fine dresses, scented flowers on the hair and beautiful ornaments on the Chest.

Chapter

2

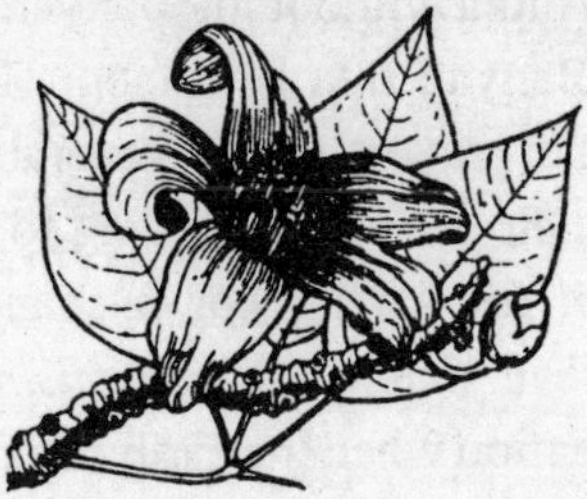

Trees of Cool Shade

"A tree that looks at heaven all day and night
And lifts her branchy arms to pray the Almightly"

2.1 Banyan tree (*Ficus beneghalensis* L)

Family: Moracaee

Vernacular names

Bengali: *Bot, Bar;* Gujarati: *Vad, Vadlo;* Hindi: *Bar, Bargad;* Kannada: *Alada Mara;* Malayalam: *Ala, Vatam.*; Marati: *Vada;* Sanskrit: *Vata Bahupada*; Tamil: *Alam, al;* Telugu: *Marri*

"It was a pleasant sight to look at the venerable tree,
Like a gaint devil with uncombed hair during dark night,
Seems to be club of evil spirits at new moon,
Appears temple with several straight columns at midday
Provide cool shelter to the men and animals,
Affords welcome relief to the travelers from the burning sun."

Legends

No village was found in India without the Banyan tree. There are so many legends about Banyan tree in ancient Hindu scripts. The childless couple usually offer prayers to this venerable tree during the month of *Jyesta* to fulfill their desire. According to *Mahabhagavata*, Savitri worshipped Banyan for saving her husband Satyavan from Yamadharma. Hindus believe that Lord *mahavishnu* sleeps on the Banyan leaf at the end of creation when the earth completely submerged under water during *Jalapralaya*. Lord Srikrishna known as '*Vatapatrasai*' as he slept on Banyan leaf during his Childhood.

Popular Lore

The name 'Banyan tree' derived from hindi word *'Baniya'* which means "merchant". In ancient days, the members of Indian merchant class used to carryout trade in persian Gulf, with pearl, diamonds and textile products. These Baniyas frequently sit under this tree to discuss about business. Hence Europeans named the tree as Banyan. Hindus usually plant the tree near the temples, shrines and burial ground and seen with respect. In villages, *Banyan* trees forms the Club-houses and rallying points because of its cool shade during hot day. Much gossip will be exchanged on politics under the Banyan tree in rural areas. The trees should not be planted near the homes and buildings. The long roots grow both vertically and horizontally and penetrate into the foundations. After few years, the walls and verandahs will be cracked.

Multipillered Immortal

Banyan is one of the gaint trees of the world. Under natural conditions the tree lives several centuries and even millenia as it is immortal. The prop roots born on the branches enter into the ground and form mighty pillars. These pillars supports the heaviness of branches. Some of the oldest venerable trees are found in Royal Botanic Garden, Sibpur near Calcutta, Adayar Theosophical Society near Chennai, Goppur in Bihar and Kampto with same state. The Banyan present in Kampto spreaded in two hectare field.

The original stem of Sibpur Banyan reached to a girth of 16m. The number of prop roots was 480, and the circumference of crown was 40m. The Banyan of *Satara in Maharashtra* is still larger and reached more than 500 metres. The spread is so surprising that one batallion of army can take shelter under the crown of tree. These trees are principal attractions to the foreign tourists.

Geographical diversity

Banyan is a native of India as it was found in ancient sanskrit literature. In Sanskrit it is called '*Bahupada*' means '*multifooted*' and *Vata* which means bearing prop roots from the branches. It was very common from Sub-Himalayan region to Kanyakumari. Rich growth was observed in moist tropical regions and adapts well in different soil conditions. It can withstand severe drought as its roots reached to ground water present in deeper soil layers. Hence the village people believe that the presence of Banyan and termite monds indicates the source of abundant ground water and dig the open wells and borewells in these sites. The trees are abundant in hill ranges of Eastern and Western Ghats.

Morphological features

It is an enormous tree and attains to a height of 30-45m and sending down roots from the branches. These prop roots enter into the ground and form trunks and thus extends the life to tree continuously. These prop roots are slender and cotton thread like at the early stage and browsed by donkeys, buffaloes and Goats. The dark green thick leaves are oval shaped with entire margin. The leathery leaves are shiny when sunshine falls on it. The tree bears very minute flowers and fruits. The inconspicuous seeds are held together in the fig or a sort of pouch. The fig is a composite fruit with large number of minute fruits. The figs are borne from the leaf stalk are green at the early stage but later turned to bright red in colour and attracts birds. The monkeys and birds like crow and green pigeon very fond of these figs and eat away. The figs are unsuitable to human consumption because of their insipid taste and fruits are infested with worms.

Menace of seed to temples, forts and buildings

The seeds of banyan tree are swallowed by birds. These birds are highly effective dispersal agents of seeds through droppings. Most of the seed is undigested in the guts of birds, germinate in the crevices of temples and old buildings. The cracks become widened due to action of roots and buildings are ruined beyond repair. No ordinary weedicide can kill these trees, except very few aboricides.

Economic utility

The figs are food for wildlife and avion population. Some tribals eat the fruits at the time of scarcity. The tender leaves are lopped for fodder to goats. The wooly tender prop roots are eaten by cattle and dried roots are used as fuel wood by washermen for boiling the clothes. The thick dry leaves are employed for making paper plates and cups.

The wood is highly suitable for making paper pulp. The wood is not used for house construction as the *Vastu* scriptures banned the wood for residential purposes. However, the aged wood is used for making yokes, well curbs, cart shafts and for general carpentery. The latex is recommended externally for joint pains, rheumatism and lumbago. The bark is a good astringent and used in case of diarrhoea, dysentery and diabetes. Leaves are cooling agents and used as a tonic in Ayurveda. The root powder is used in tooth powders to arrest the tooth decay.

Place in bioaesthatic Planning

The tree is best suitable to plant along the National high ways, road sides and in the grazing lands of dryland tracks, gives cool shade to humans and animals alike. It can be planted around the bird sancturies as it provide shelter to several birds.

2.2 Bael tree (*Aegle marmelos Correa* ex Roxb)

Family: **Rutaceae**

Vernacular names

English: *Bengal Quince*; Gujarati: *Bil;* Hindi, Bengali & Marati: *Bel*; Kannada: *Bilpatre;* Sanskrit: *Bilva;* Tamil & Malayalam: *Vilvam;* Telugu: *Maredu*

"Tridalam, trigunakaram, trinetram cha triyayudham trijanma papa samharam, Ekabilyam sivarpanam"

-Shiva Purana

"The Lord who is having three gunas, three eyes, three weapons, destroyer of sins for three births. I dedicate the bilva leaf having three leaflets to that supreme."

Popular Legend

The Bael tree is known to Indians since vedic period (800 BC). Several references have been made in *Yajur Veda* and *Mahabharata* about value of the tree. According to Hindu mythology, the tree is another form of Lord *Kailasanath*. The leaves are traditionally offered for the worship of Lord Shiva during the month of *Kartika*. The trees are planted around the hindu temples due to its mythological significance.

Geographical diversity

Bael tree is indigenous to Indian subcontinent and mostly found in Tropical and subtropical regions. The tree also found in wild in lower ranges of Himalayas upto an elevation of 1200 metres. The tree is also occurs in Pakistan, Bangladesh, Sri Lanka, Myanmar, Thailand and Vietnam.

Morphological features

Bael tree is a medium sized tree which lives and bears fruits for a long time. Some trees of more than 80 years age, bears fruits heavily. The deciduous tree is glabrous and shed its leaves during summer. New leaves emerges immediately. Leaves usually grouped into 3 and rarely 5 also. People believe that the leaves are another form of *trinetra* (3 eyed God) and highly sacred. The green leaves resemble litchi and similar to mango leaves. Flowering starts after the tree putforth new leaves. Flowers are fragrant and contains sweet honey. The leaflets ovate-lanceolate, acuminate and dotted with glands. Inflorescence is small lateral panicle and have greenish-white flowers. The sweet scent emits from flowers attracts honey bees. Fruit is globose, 6-18cm diameter, with smooth grey woody shell. The fruit ripens after one year in the month of March. The fruit contains a mass of orange coloured, sweet, mealy pulp.

Economic Utility

Unripe fruits are astringent, digestive and stomachic in Ayurvedic system. The fruits are good remedy for diarrhoea and dysentery. Pulp is aromatic and cooling when used in the form of *sherbet*. *Marmelosin* is the active constituent in the fruit pulp. It acts as a laxative and diuretic. It is cardiac depressant when used in strong dose. Gummy substance in young fruits. It is also used as a varnish for pictures and adds brilliancy to water-colour paints.

Dried fruits, freed from pulp, used as snuff ball and pill boxes. Stem yields a gum used in book binding. Leaves are lopped for fodder. The infusion of root bark is used to cure fevers by *bastar* tribals of Madya Pradesh. The wood is used for making small house hold articles and cattle sheds.

Place in bioaesthatic planning

The trees can be planted as avenues in University campuses, Parks, hospitals and other public places.

2.3 Black Siris (*Albizzia odoratissima* Benth)

Family: Leguminosae – Mimosease

Vernacular names

Bangali: *Kakursiris;* Gujarati: *Kalo sarasio;* Hindi: *Kalisiris;* Kannada: *Bilvara;* Malayalam: *Pulivaga*; Marati: *Chikunda;* Tamil: *Kasuvagei;* Telugu: *Chinduga*

Geographical diversity

The trees has no preference of soil but luxurious growth can be seen in fertile soils coupled with good drainage.

Morphological features

Black siris is fairly large sized tree which appears dark in colour due to dark-grey coloured trunk. Bark is dark-grey with patches of darken colour with grey tomentos. Leaf rachis is 10-25 cm long, down with a gland near the base. Rachis has 3-8 pairs of Pinnae. Each pinna has 8-10 pairs of leaflets which are linear-oblong, unequal-sided, acute and minutely mucronate. The leaflets are dark green in colour with glaucous beneath. Inflorescence is a terminal panicle, very rarely auxillary with Cormbose branches. Flowers are pale white, sessile and fragrant. Pods contain 8-12 seeds with reddist-brown and some times dusty-green in colour. Seed position can be identified with darken blotches on pod. It flowers during summer, fruiting takes place in January and February.

Economic utility

The stem consists of large white sapwood and dark brown heartwood. The wood is very hard, fairly durable and seasons well. As it takes fine polish and presence of dark streaks, the wood is highly suitable for house furniture. White siris wood is used for crushers of oil mills, wheels, butter churners, handy sugarcane crushers and household furniture and panelling. The leaves and twigs are lopped for cattle and camel fodder. Bark is the source of a fermented drink, called *Basi* which is used in Philippines. The tree yields a gum which is used for book binding.

Place in bioaesthatic planning

Black siris can be planted as avenues in national highways, industrial sites and along the canal banks.

2.4 Heaven tree (*Ailanthus excelsa*, Roxb)

Family: Simaroubaceae

Vernacular names

Gujarati: *Ardusi;* Hindi: *Maharuk;* Kannada: *Doddamara;* Malayalam: *Mettigongilyam;* Marati: *Mahanimb;* Oriya: *Mahala;* Sanskrit: *Madala*; Tamil: *Perumaram;* Telugu: *Peddamanu, Swargeloka chettu*

Geographical diversity

Heaven tree is said to be a native of Australia. William Roxburgh found this tree in Circar districts of Eastern Coast of India during 18th Century and claimed that it was indigenous to India. This tree is most commonly found in drier parts of Gujarat, Bihar, Central India, Deccan Plateau and Penisular India. Heaven tree prefers impoverished soil and very high temperatures. Another species *Ailanthus malabarica* DC, was more common in western Ghats, Myanmar and Sri Lanka. The aromatic resin oozed from the wounds of bark is used in the treatment of dysentery and making *agarbatties*. In Kanara district of Karnataka, abundant stands can be seen. Other spees *A. glandulosa* is mostly found in Indo-Gangetic region as shade tree. It is believed to be native of China and various parts used extensively in Country medicine.

Morphological features

Heaven tree is a fast growing deciduous tree growing a height of 25m, having white bark. Some trees reaches to a height of more than 30m. in

Western coast. The leaves are pinnate and 30-80cm long. Leaflets (6-12 Pairs) are arranged in alternate manner on pinna. Leaflets are variable in size with toothed margins. Inflorescence is panicle and much branched. Flowers are creamy/Yellowish and about 10mm across. Fruit is a flat pod, lance shaped, sharp at both ends. The pod contains only one seed.

Economic utility

The bark of the tree is used as a tonic and febrifuge in native medicine. The bark resin has medicinal properties and used for curing dysentery. The wood is light in weight and pale and specially used for making sword handles. In coastal areas, the wood is employed for making fishing boats, packing cases, farm implements, toys, matches, veneers and ply-wood. The bark extraction is used as anthelmintic, expectorant and antispasmodic. It is good substitute for *kurchi* bark. Tree yields hog gum or *Bassora* which is used for binding books. The tree serves as host for *Eri silkworms* Shoots are lopped for cattle fodder.

Place in bioaesthatic planning

The tree can be planted as avenues in national highways, village market places and cattle yards. It is a good shade giving tree.

2.5 Jackfruit tree (*Artocarpus heterophyllus*, Linn)

Family: *Moraceae*

Vernacular names

Bengal: *Kanthal;* Hindi: *Kathal;* Kannada: *Halasu;* Malayalam: *Chakka;* Marati: *Phanas;* Sanskrit & Telugu: *Panasa;*Tamil: *Pilapalam*

Jack tree enjoys special favour in home gardens on the West Coast and is also used as one of the shade trees in Cardamom, Pepper, areca and Coffee plantations in South India. The fruits are extensively produced in the back-yards as well as amidst the plantation crops but the demand is meagre in the outside regions. There is a wide spread belief that consumption of Jackfruit leads to certain digestive ailments.

Geographical diversity

Jack tree flourishes best under humid and warm climate of hill slopes to an altitude of 1530m. If moisture is not a constraint, it can also be grown in arid and warmer plants. Soil drainage is most important because of sudden decline of numerous trees, in areas suffering from a sudden rise of water table. Jack tree prefers a deep, rich and well drained alluvial soil or very open-textured lateritic soil, if it is supplied with adequate nutrients and water. Usually the trees grown above 120m altitude produce the fruits of inferior quality.

Morphological features

Jack is a large evergreen tree of 25-30. height, and all parts are filled with latex. The leaves are evergreen, fleshy, leathery, coriaceous, simple, alternate and stipules leave a sear into branch. The inflorescence is syeonus and male flowers present above the female flowers, on the inner side of fleshy, receptacle. Flowers are small, Unisexual and regular. The perianth is persistent. Fruit is a sorosis and multiple. The fruits are borne on the trunk and main branches. They are oval, 60cm length, 30cm in diametre and weigh as much as 40 kg. The skin is studded with short spikes, pale green at young becomes yellowish when ripen. The fruits contain a yellowish juicy sheath with strong flavour in which large number of seeds are embedded. The *Kapa* variety has a sweet, fleshy and crisp pericarp which is considered to be better than *burka* variety which is sour.

Economic utility

Apart from its use as a table fruit, jack fruit is a popular with the house wife from making pickle, for dehydration into jackfruit leather or papad and other forms of culinary preparations. The fruits are made into syrup, Jam, Jelly, Candy and canned with sugar and honey. Salted flakes are used as a substitute for potato chips, after frying in ghee. The outer pericarp is a prized cattle feed, and relished by the cattle when mixed with rice gruel. The seeds are roasted and eaten. The latex from the bark contains resins and often used to plug holes in earthenware or buckets. The timber is prized for making furniture. The leaves are relished by Goats. They are also stitched together for use as a dining plates. The fruits are rich in vitain A and C. The tree of such diverse value deserves to be grown widely to promote its economic worth. The wood is also used for making musical instruments like Veena.

Place in bioaesthatic planning

The tree can be grown as avenues on road sides, as shade trees, car parking places and in public parks. The true is known as excellent shady tree in East Cost.

2.6 Lucky Bean tree (*Putranjiva roxburghii* wall)

Family: Euphorbiaceae

Vernacular names

Bengali: *Jioysuta;* Gujarati: *Jivputrak;* Hindi: *Putranjiva;* Marati: *Patravanti*

Geographical diversity

Luckybean is a handsome evergreen tree and distributed from sub-Himalayan tracts, to Kanyakumari, through Indian peninsula. It was also found in Sri Lanka. Abundant stands can be seen along the Western Ghats. In other parts of India, Putranjiva is a scarce plant. The tree prefers moist and shady regions. Dense stands of luckbean trees appears on the banks of Suswa river and swamps of Rishikesh and Khairi. The botanical name *Putranjiva* derived from sanskrit word *Putra* means "son" and *jiva* means "life". The specific name *roxburghii* refers to william Roxburgh, the first Superindent of the Botanic garden of Calcutta.

Morphological features

Lucky bean is evergreen elegant tree and reaches to a height of 25m, Under ideal conditions. The erect trunk consists of several pendant branches. The crown is spreading with innumerable expanding branchlets which gives deep shade. The tree is smooth except the young branches which contains hairs. The leaves are leathery, shiny and shaped like lance head. New Borne leaves have a very pleasant parrot green colour and become darken and shiny after one month. Flowers are small, unisexual and yellowish in colour. The male flower is short pedicelled, in axillary clusters which often spicate. The females are long-pedicelled, axillary, two or three together. The tree produce new leaves in April and flowering occurs in hot summer. The development of fruits takes place during winter season. The fruit is drupe, Oval shaped, fleshy 14-19cm long, enclosing a very hard and rough surfaced stone.

Economic utility

The leaf juice is used as medicine as a relief from colds and coughs. The extracted oil from seeds used for burning in tribal areas. The green leaves are lopped for cattle fodder. Rosaries made from the nuts are often used by fakirs, sadhus and devout people for the protection against the evil eye. The nuts are strung up in chains and tied round the neck of child to keep off diseases. Pregnant women also wear the rosary to prevent abortion.

Place in bioaesthatic planning

The tree has a high reputation of giving cool and pleasant shade and can be used as avenue in all tropical countries where sun burning is severe. This graceful tree can be planted along the village, town and city roads which gives relief to the travellers. Finest avenues can be seen in Indian Botanical garden of culcutta. The trees are suitable to plant around hospitals, courts and school and college premises to beautify the surroundings.

2.7 Mahua or Butter tree (*Madhuca indica* J.F. Gmel)

Family: Sapotaceae

Vernacular names

Bengali: *Maul Mahwa, mahula;* Gujarati: *Mahuda;* Hindi: *Mohwa, manwa;* Kannada: *Hippe;* Malayalam: *Poonam;* Marati: *Mahwa, mohwra;* Oriya: *madgi, mahoa;* Sanskrit: *Madhuka*; Tamil: *Illupei;* Telugu: *Vippa*

Sumptious dinner to Wild Animals

Any literature on Indian trees would not complete if the Mahua tree was excluded. It is highly useful to both humans and wildlife. If one goes into deeper forests during March and April, he can observe the crowd of deers, wild boars and Jackals around the tree, busy in consuming the fleshy flowers dropped in the morning. Similarly crowded children can be seen in the Mahua avenues of villages, busy in collecting the sweet corollas. The collected corollas are sold to liquor makers who manufacture country liqueors and vinegar. The author also experienced the drops of flowers like rain drops when he moved through the Mahua avenues on morning walk while he studying at Banaras Hindu University. The bees also very busy for collection of nectare from the flowers found on the tip of branches. It is a beautiful experience of continuous flower drop coupled with buzzing sound of honey bees on the top and strong floral smell while walking in the middle of mahua trees at dawn.

In the central India, sloth bears are very fond of flowers and eat them in huge quantities. This sumptious dinner make them intoxicating and sleepy. The flowers ferment in the stomach and alcohal will be released, consequently these animals behave like drumkards. The *Gonds, Koyas, Banjara,* and other tribals prepare country liquors and enjoy them on festive occasions. Tribals loose temperament when they swallow large quantities of this country arak. It is said that during Maratha and British rule, most of the mahua trees are cutdown to arrest the illicit distillation of country arrack and to stop the lawlessness of tribals during Mahua flowering. The dried flowers are used to give as *Prasada* to devotees at Srirama temple of Bhadrachalam located in Eastern Ghats.

Geographical diversity

Mahua tree is a native of Central India. It mostly distributed in Western Ghats, Eastern Ghats, Gujarat and Chota Nagpur of Bihar. It can also be seen in sub-Himalayan region of Punjab, and Himachal Pradesh. Mahua was found from plains to 900m. altitude. Mahua trees are planted extensively in South Indian States.

Morphological features

Mahua is a large deciduous tree reaching to a height of above 20m. in fertile alluvials. The branches are well spread and had large crown. The leave are elliptical, hairy when young and become glabrous later. The leaves are borne at the tip of branches as a cluster. The tree flowers at the end of winter (March-April) and fruiting occurs in rainy season (June to August). Flowers borne as dense fascicles near the end of branches. The tree shed its leaves before flowering. Flowers have long stalks, droopy, rusty and pube cent. The fruit is fleshy ovoid in shape, green at initial and turned to reddish later.

Economic utility

The flowers are fleshy and rich in sugars and used for making syrups, Which produce coolness during summer. Flowers are demulcent and used as tonic for general weakness. The flowers are fermented and distilled for making liquors and vinegar. Corolla is eaten raw and as dried. The stamens are boiled for about 6 hours, left to simmer until the water evaporates. The softed stamens are eaten with rice and tamarind. These are cooked with rice, made into sweet meats. Children eat the dried ones along with parched grains. A favourite liquor is distilled from corollas. One gallon of liquor can be distilled from 100kg of dried flowers. This liquor contains about 95% alcohol. Syrup can be made extracted from corollas. Jams can be prepared from the flowers. Mahua berries are eaten after cooking. It is used as a substitute for honey.

The wood is very hard and employed for making carriages, sports goods, muscical instruments, and oil presses and sugarcane crushers. The wood is also used for ship building.

Place in bioaesthatic planning

The trees can be planted as avenues along the town and city roads, national highways, picnic spots and in University campuses.

2.8 Neem (*Azadirchta indica* A. juss)

Family: Meliaceae

Vernacular names

Bengali: *Nem;* Gujarati: *timba;* Hindi: *Nim;* Kannada: *beru;* Malayalam :*Vepe;* Marati: *Limba, nimbay;* Persian: *Azadirachta;* Punjabi: *neem;* Sanskrit: *Nimba;* Tamil: *Vempu;* Telugu: *Vepa, yepa*

Eco-morphological features

There is no doubt about the origin of neem from India as its botanical name itself indicating *Azadirchta indica*. The word *indica* means Indian origin. Some people confuse between Indian neem (*A. indica*) and *Melia azadirachta* L. as the morphological features are the same. The only difference lies in the colour of inflorescens. The flowers are white in case of Indian neem and lilac in persian neem. The plant is well distributed throughout the Indian subcontinent in dry and Semi-dry regions. It is sensitive to frost and waterlogging.

Neem is a medium to large sized elegant tree with a straight trunk. The tree is ever green, dark green leaved and has dense round crown. Leaves are compound, long, pendent and crowed appearance at the apex of branches. Young leaves are pale green, thin and tinted with rusty spots. The tree shed its leaves in March but a little period. The flowers are greenish white and lightly sweet scented. Blossoms appear in the first week of April during hot season.

As it is having more leaf area index, its photosynthetic rate is high and produce enormous quantity of oxygen and cool shade. According to legend, *Sadguru Sainath* meditated several years under neem tree during his childhood. Even now the scripture written near *Gurustan* of Shirdi indicating its sacredness.

> ***"Sada nimba Vrukshya, Mooladhi vasat***
> ***Sudha sravinam tikthamasya priyantam***
> ***Tarum Kalpavrikhadhikam, sadhayantam***
> ***Nama meeswaram, Sadgurum Sainatham"***

that is

> ***"Owing to continuous sitting under the Neem tree, the tree lost its bitterness and converted its sap into sweet ambrose. I bow down to the Sadguru Sainath, Universal God who made the neem into wish yielding Kalpavriksha."***

During the hot summer, the sap oozes from the stem which the phenomena called *Neem weeping*. The collected sap is used as tonic.

Propagation

Neem is mainly propagated through enormous quantity seed produced during the onset of monsoon. The tree is considered as hard as it stands pollarding, saline and severe drought conditions.

Economic utility

All most all parts of the tree find various uses in medicine, construction and fodder. The timber is termite proof and employed for making panels, boards, *desi* ploughs, toys and other house building purposes. The amber coloured gum and neem toddy is used as a tonic and other alimentary canal troubles. The Margosa oil obtained from seed is having garlic like flavour and highly effective in the treatment of leprosy and skin diseases.

Nimbidin is the main bitter principle of oil which can be utilized as bactericide and insecticide. Bark decoction are autiseptic and applied

Neem oil is effective in curing mange in dogs, to boils and eczeme in the form of poultice. Flowers are stomachic tonic and extensively used in Ayurvedic medicines.

Dried leaves are placed in books, stored bins of pulses and cereals for keeping away the moths and stored grain pests. Neem cake can be applied to control citrus canker and leaf miner. Odours of burnt leaves kills insects and mosquitoes.

Place in bioaesthatic planning

The trees can be planted on both sides of national high ways as they give good shade to the travellors and vehicles. Spacious streets of cities and metropolis can be made with beautiful avenues of neem trees. In case of huge trees, the crown can be cut off from top of the stem during winter. New shoots are borne in March and gives delight to the viewers. The seeds are not having long viability and hence sown soon at the onset of monsoon.

2.9 Pongam (*Pongamia pinnata* (L) Pierre)

Family: Papilionaceae (Fabaceae)

Vernacular names

Bengali: *Kanj Papre;* Hindi: *Karanj, paper;* Malayalam: *Unne;* Marati: *karanj*; Tamil: *Pongam, Punku;* Telugu: *Kanuge, Pungu*

Geographical diversity

In western Indian states the tree goes under the name of Karanj. The botanical name *Pongamia* derived from Tamil name '*Pongam*' and species name 'Pinnate' means "having pinnate leaves". Pongam is indigenous tree of India and grown throughout the country. It was distributed from the foot-hills of Himalayas down to the south of Peninsula especially near the sea-coasts. The tree was found luxurious growth along the river and stream banks, pond bunds, canals and also in dense, deep forests. In some forests, where rainfall. is abundant, the tree attains to a height of 30m. It cannot tolerate ill drainage and prefers sandy grounds in the neighbourhood of brooks, canals and seas. Dense population can be seen in the forests of Madhya Bharat.

Morphological features

Pongam is a dwarf deciduous tree (10m) with a short bole and umbrella like dense crown which gives cool shade. The bark is smooth and grey in colour. Under favourable, conditions, the tree attains to height of more than 20m. The leaves are glossy and bright green. The compound leaves consisting of 5-9 leaflets, oval in shape, opposite in pairs with an elongated tip. Often the leaves are marked with ugly vein-lines due to damage of nematodes and insects. The tree shed its leaves at the end of winter. The inflorescence is raceme produced in the axil of leaves. The raceme is 20cm long covered with purplish or lilac flowers. Variation in colour also seen in various places. Some flowers are pure snow white and another flowers are white but with a touch of lilac on the wings. As a whole, at the full bloom the tree branches covered with rich purple colour. The tree is at full bloom in the hot months of April, May and June. The tree become bare and leafless before the flowers cameout. This attracts the attention of the public. Fruits are woody pods and develops showly and reached to maturity at the beginning of monsoon in the following year. Pods are 5cm long, compressed and narrowed at both tips.

Economic utility

The leaves are used as green manure in rice to drive away white-ants. It is a best shade tree which gives cool shade during hot summer months. Oil is extracted from the seeds and used for the country lights by tribals. Oil is also popular medicine, applied to rheumatic pains and skin infections. The deoiled cake is used as concentrated organic manure.

The wood is not hard and used as firewood in rural areas.

Place in bioaesthatic planning

The trees are highly suitable for road-side avenues, car parking places, railway station, platforms, city Bus station, Aerodromes. It is realy worth when the tree is laden with charming mouve flowers.

2.10 Rosewood or Bombay blackwood (*Dalbergia latifolia* Roxb)

Family: Fabaceae

Vernacular name

Bengali: *Sitral;* English: *Bombay rosewood;* Hindi: *Shisham;* Kannada: *biti;* Marathi: *Shisham;* Tamil: *Iti, eriwadi;* Telugu: *Nalla virugudu cheva*

Rose wood is one of the most highly prized and priced timber trees of the world. The wood is very dark purplish brown, beautiful figured with black streaks. The generic name *"Dalbergia"* named after N.Dalberg, a Swedish botanist of early 19th century. The species name *"latifolia"* means 'having leaves'. A close cousin of rosewood, *Dalbergia sissoo*, grows extensively along the rivers of Himalayan and Vindyan originated rivers. It's wood is still dark and its wood is highly priced and employed for making top class cabinets and bed room furniture.

Geographical diversity

Rosewood is a native of India and distributed from the Himalayan foothills to Peninsular India. In mountain tracts it is found to grow upto an elevation of 1500m. The tree is small and crooked in dry deciduous forests of central and western India and reaches upto 10m. only. In moist peninsular India where rainfall is abundant, large trees of over 35m. high also can be seen in Southern reaches of Western Ghats. It does well in Well-drained moist soils.

Morphological features

Rosewood is a medium to large sized tree having spreading branches forming a shady canopy. The tree is deciduous in dry areas but some leaves are present throughout moist regions. Leaf-shedding usually occurs at the end of cold season and at the start of spring. The compound leaf consists of 5-7 leaflets alternately arranged on central stalk. The leaflets are orbiscular, firm, green on the upperside and pale beneath. The flowers are white, similar to Pongamia, arranged in lax cluster inflorescence from leaf axils. The fruit is flat pod, strap shaped having 1 to 3 seeds. Flowering occurs in hot summers and fruits are formed in rainy season and remain on the tree for some months until the next rainy season.

Economic utility

The tree gives most valuable timber which fetch highest price. The wood is strong, very hard, close grained, purple black in colour but heavy. It takes a beautiful polish and excellent furniture can be made.

Place in bioaesthatic planning

The tree can be planted along road sides which gives elegance to roadview when it is full of green leaves.

2.11 Tamarind (*Tamarindus indica* L)

"Through the Tamarind tree dies but its sour can not be killed"

– Telugu maxim.

Family: Caesalpiniaceae

Vernacular names

Bengali: *Amli, ambli;* Hindi: *Imli;* Marati: *Chinch;* Tamil: *Puli;* Telugu: *Chinta*

Superstitious awe

Though tamarind fruit and juice is used in every curry and dishes in South India, the tree is considered as *superstitious awe* by rural folk. People do not see the tree in the early morning because they believe that it was ill-omened. During summer nights, rural folk seldom sleep under the shade of tamarind because of its excessive carbon dioxide production during respiration. According to folk stories, the tree is an abode of malevolent spirits and devils and that is why it was not seen near the houses and dwellings except abandoned old houses. In ancient days, tamarind trees were avoided for camping the army or revenue staff, as the acid of fallen leaves damage the cloths and tents. Surprizingly no plant can be seen survive under the shade of tamarind possibly due to its high acid nature of leaves which inhibit the growth of other plant communities. Perhaps on account of acid nature, people believe that they get nasal and lung diseases if slept under the tamarind tree for a night.

Geographical diversity

The word, *Tamarind*, derives from persian language, *Tamar-i-Hind* which means an Indian date. The specific name 'indica' indicates the Indian origin. The persian merchants thought its origin is Indian subcontinent. In fact, the original home of the tree is tropical Africa. Now a days it was cultivating throughout Asia, Africa and America continents and well grown trees can be seen in eastern uttar pradesh, particularly in Ayodya region. Although luxurious growth was seen in tropical and subtropical regions where rainfall is medium to high, the tree can withstand drought excellently.

Morphological features

Tamarind is a handsome, evergreen tree with large spreading crown. The bark is rough, dark-grey in colour. The trunk is straight in young trees but irregularly thickened in older trees. Some times big holes or caves are seen in large trees when the trees are damaged during storms.

The timber is very hard and the branches are not broken even they carry huge weights. The leaves are compound, consisting of 10-20 pairs of small leaflets. They are grown very densely on the tree and cast an unbroken and cool shade on the ground. Young leaves are shiny and light green. Old ones are darkgreen but loose their brilliancy. The newly emerged leaves are used as green vegetable. Flowers appear in hot months, from April to June. The inflorescence is raceme and borne at the end of branches. The flowers are inconspicuous and the petals are variegated yellow and red. Two petals are reduced into scales. A dense and yellow carpet of flowers can be seen on the ground beneath the tamarind during flowering season. The pod is curved like sickle and irregularly swollen due to presence of seeds. These seeds (3-12) immersed in fibrous pulp. The acidic pulp is used in curries and pickles. Some varieties have a reddish pulp with sweet-sour, ripen during November and December. The acid pulp is most important article of Commerce in South India as it used as condiment in curries.

Java Beer with pleasant flavour

Javanese prepare a superior drink like *Dutch beer* from tamarind pulp by mixing with black Jaggery. They take river water and add Jaggery, tamarind pulp and few lemons and ferment the mixture for about one day in wooden vats. After fermentation they boil the liquid to expel impurities and dirt to the surface. This liquid has delicate flavour like Ducth beer in Holland. This beer cools the body and improves the digestion. People drink several glasses as it was good for health in hot tropical climates.

Economic utility

The fruit pulp is extensively used as souring agent in curries, sauces and chutneys. The pulp is refrigerant. carminative, laxative and excellent in curing for biliousness and febrile conditions. Being antiscorbutic, the pulp is utilized by sailors instead of lemon juice. The kernels are used as food for tribals at the time of scarcity. The kernel powder is an excellent sizing material in textile industry. The jellose of powder is used for the formation of gels and a substitute for fruit pectins. It is also adhesive in book-binding, cardboard and leather industries. The seed oil is one of the ingredient in paints and varnishes.

The wood is employed for making rafters, beams, trusses and purlins. As the wood is very hard, rice pounders, mallets, printing blocks, oil crushers, wheels and other agril, implements are made extensively. The wood also yields decorative plants for paneling and furniture.

The flowers are good source of honey, however, it was slightly acidic in taste. The tender leaves and flowers are used as vegetable. Tamarind is used to cure "flourosis", deformation of bones and teeth due to drinking fluorine rich water.

2.12 Iron wood tree (*Mesua ferea* L)

Family: clusiacea

Vernacular names

Andaman: *Gangane;* Assamme: *Nahor;* Gujarat & Marati: *Nagchampa;* Hindu & Bengali: *Nagkesar nagesar;* Kannada: *Nagakesara;* Malayalam: *Nanga;* Sanskrit: *Nagakeshara;* Tamil: *Nangu, Nagachambagam;* Telugu: *Nagakesara;* Trade: *Mesua*

Geographical diversity

Iron wood tree found in Southern and Northern tropical wet evergreen forest of Assam, Meghalaya, Nagaland, Arunachal Pradesh, West Bengal, Kerala, Karnataka, Bihar and Andman. It is usually grows in the warm, moist and equatorial habitat where the rainfall ranging from 2000-5000 mm. Mesua prefers deep, fertile – and well drained deep Sandy loams but come up on moderate heavy tools. The tree regenerates profusely in its natural habitat because of its abundant seed production.

Morphological features

Mesua is a medium to large sized, evergreen and elegant tree attains to a height of 45 m and with a girth of 3 m. The tree buttressed at the base and spirally twisted and has a dens crown. Leaves lancolate or elliptic – oblong, thin, glabrous and shiny on upper surface. The leaves are persistent and falls after two years. Flowers are white, highly fragrant and solitary. Fruit is ovoid, 5 cm in diameters and contains 1-4 seeds. The old leaves fall during winter and new flush appears at the end of winter. The plant produces fertile seed from 15 days onwards.

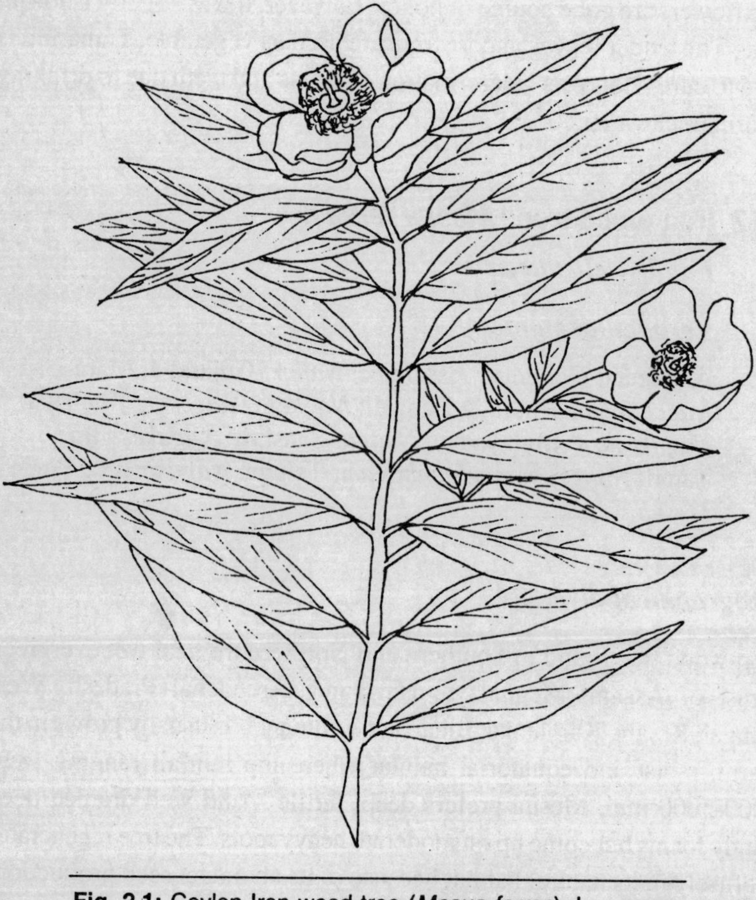

Fig. 2.1: Ceylon Iron wood tree (*Mesua ferrea*), L – green, Fl – white (yellow stamens).

Economic utility

Mesua is one of the hardest and strongest of Indian timbers. Being extremely hard it is very difficult to saw even in green conditions. The wood is used for railway sleepers, bridges and posts, beams, electric poles, carrier's cutting blocks, boat building, crushers, bearing tool handles, golf club heads and walking stick. As it is highly durable it is also used as bridges, camel locks where strongness is needed. It is also used for making axles, felloes, spokes and stocks of country carts.

Seeds yield a fatty oil which may be blanched and used for soap making and also used for skin troubles. It is also good for rheumatism when applied externally. Flowers used for cough and buds eaten for dysentery. Bark along with Ginger used as sudorific. Pillows are filled with stamens to impart a pleasant odour in Malaysia.

Place in bioaesthetic planning

Mesua is cultivated as an ornamental tree in gardens and avenues in North-eastern states. These can be planted on the road – sides of national highways, industrial sites, railway stations, and Bus stations.

wood is used for railway sleepers, bridges and posts, cement, electric poles, carts, cutting blocks, boat building, furniture, bearing, tool handles, golf club heads and [illegible] stems. As a telegraph [illegible] it is also used as bridges, camel locks where strongness is needed. It is also used for making axles, felloes, spokes and stocks of country carts.

Seeds yield [illegible] oil which may be bleached and used for soap making and may used for skin troubles. It is also good for rheumatism when applied externally. Flower used for cough and buds used for dysentery. Bark along with ginger used as febrifuge. Pillows are filled with [illegible] to impart a pleasant odour in Malaysia.

Place in [illegible] planting

[illegible] is cultivated as an ornamental tree in gardens and avenues in North-eastern states. These can be planted on the road sides of national highways, monument sites, railway stations, and bus station.

Chapter

3

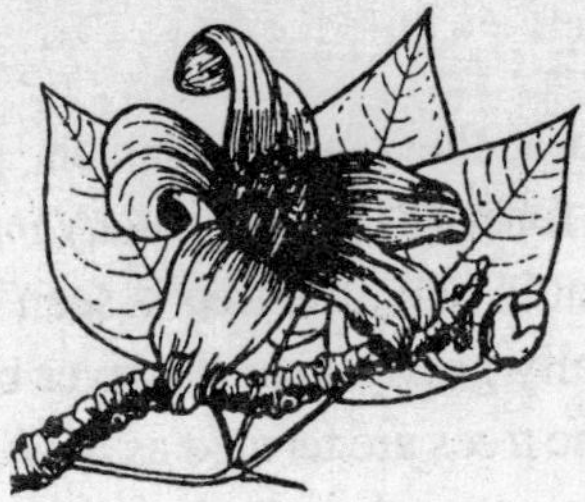

Trees of Gracy Foliage

"The tall, majestic tree lying its feet in the earth
Erects its head in the sky
For watching the brown clouds passes by
Drink the sun rays and gives cool shade to animal world"

3.1 Arjun (*Terminalia arjuna* (Roxb) Weight & Arn

Family: Combretaceae

Vernacular names

Assame: *orjun;* Bengali: *Arjhan;* Gujarati: *Sadado;* Kannada: *Maddi;* Marati: *Sanmadat, Sadaru, Vellamarda;* Oriya: *Arjuno, Sahajo;* Punjabi: *Arjan;* Sanskrit & Hindi: Arjuna; Tamil: *Vellamatta;* Telugu: *Yerramaddi;* Trade: *Arjun*

Legend

Like his father Devendra, the king of heaven, people believe that Arjuna tree is another form of *Pandava madyama*, the great hero of Mahabharata war. That is why, the tree is seen with respect by the hindus. The trees are usually grown near the river banks and in dampy soils. In sanskrit scripts, the trees are termed as *Nadisarjal* which means "plants like to grow near the rivers". In latin, the tree is called *Terminalia arjuna*, the '*Terminalia*' means flowers appearing terminally and '*arjuna*' derived from hindi word which means *white body*. As the wood is white in colour like Arjuna, it is called as arjun tree.

Geographical diversity

The tree is found throughout Indian subcontinent, Myanmar and Sri Lanka particularly in tropical moist deciduous and dry deciduous forests. It is grown on various types of soils from plains to 1500m altitude. Mostly seen on the banks of rivers, brooks and streams where moisture is not a limiting factor. The trees drop their leaves but not ever seen as fully leafless.

Morphological features

Arjun is a beautiful, ever green tree and attains to a height of 25m. The bark is smooth grey, tinged with green and red. The trunk is buttressed and some times the girth may reach upto 3m. It will shed its leaves at the end of winter and new leaves are borne in April before blooming. Before the shedding of leaves, the leaves turned into copper colour and appears highly attractive.

The leaves are thick, oblong, opposite and gives good shade. The leaves on the lower surface usually carry a pair of conspicuous glands close to the top of leaf stalk. The flowers are pale-yellowish-white, cup-shaped, similar to myrobalan. The flowers are crowded on a long axis. Flowering occurs from March-June, depends upon climates. As the flowers are full of honey and attracts swarms of bees. The fruit is a nut which have leathery wings. The fruit is tan-coloured when dried.

Economic utility

The tree yields very good quality of timber which is lustrous brown and moderately heavy. The wood is used in making boats, carts, posts, beams, Yokes, railway carriages, furniture and cabinet work. Ships are mostly made with this wood. It is used for building houses, electric poles, masts and water taps.

Bark is used for tanning the hides. Bark is also employed as astringent, tonic and remedy for ulcers. Tender twigs are used for mouth washing in case of tongue blisters. The gum is also used as medicine. Bark powder is employed as febrifuge, styptic, and antidysenteric. Pulverized bark gives relief in hypertension and acts as a diuretic, in cirrhosis of liver. Fruits are deobstruent and tonic. Leaf juice is used in earache. The leaves are used as cattle fodder and contains 10% protein.

Place in bioaethatic planning

As the tree is very beautiful and gives excellent shade, it can be planted on roadsides in cities, towns and along the national highways. It is finest avenue tree and gives lush green paint to surroundings.

3.2 Ashoka (*Saraca indica* Linn)

Family: Caesalpinioideae

Vernacular names

Bengali: *asok*; Gujarati: *ashopalava*; Hindi: *Sita Ashoka*; Kannada: *ashoka*; Malayalam: *Hemapushpam*; Oriya: *Osoka*; Sanskrit: *Ashok*; Tamil: *asogam*; Telugu: *Asokamu*

Tales from Ancient scripts

Ashoka is one of the prettiest trees of world which gives thick shade and coolness to the road travellers. The trees was described in *Sundarakanda* of Ramayan written by the great sage valmiki. The demon king *Ravanasura* stolen *sitadevi* and kept in *Ashoka vana* of lanka for several months. When *veer hanuman* entered into the garden in search of Sita, he found her under the shade of Ashoka tree.

The Hindus regarded the tree as highly sacred and believe that its presence bring the wealth. According to Hindu mythology, Ashoka was dedicated to *Manmada* who is God of love and elevate the emotions of newly married couple. Buddists believe that Goutama Buddha was born under an Ashoka and considered as highly sacred. The tree was termed as symbol of fertility and was extensively sculptured in Buddhist temples of *Sanchi, Kushan Madhura,* and *Bharhut.* The trees are extensively planted around the monasteries and in the gardens by Buddhists in Myanmar and Thailand.

Geographical diversity

The tree is believed to be native of India and extensively found in Malayan Peninsula, Myanmar, Sri Lanka and Bangladesh. In India, it is commonly found in Khasi hills of Assam, hill areas of west Bengal, Western Ghats of Maharashtra and northern circars. The tree was found wild in evergreen forests of North eastern area.

Morphological features

The tree is very handsome, medium sized, ever green, and has erect habit. The branches spreads in all directions and form dense canopy and shapely crown which obstructs sunrays to reach the ground. The stem is covered with smooth, dark brown bark. In old trees, the bark turns into grayish-brown. The leaves are shiny and dark green. Each leaf is a 25 cm long, 3-7 pairs of leaflets. These leaflets have wavy edged. Young leaves are droopy, coppery red and flaccid. The leaves grow alternately on the branches. The flowers are large, compact, orange-red in colour forms into clusters. The flowers with a deep green foliage back ground appears stunning beauty to the viewers. It is propagated through seed. It flowers profusely in February and March.

Economic utility

The decoction of bark is used in uterine affections. The bark is astringent and has stimulating effect on endometrium and overian tissues and is useful in menorhagi due to uterine fibroids. It also good medicine for leucorrhoea, internal bleeding, hemorrhoides and hemorrhagic dysentery. Flowers used as uterine tonic and useful in beliousness and dysentery. Fruits chewed as a substitute for areca nut. Pods make a good forage for cattle. Wood used for making ploughs and shafts.

Place in bioaesthatic planning:

It is cultivated in the garden and public parks of Calcutta and Lucknow *etc*. It should be planted in open places as it spreads in all directions. Ashoka is most suitable tree for parks,public gardens, village and town Junctions, car parking places and university roads.

3.3 Chatiyan (*Alstonia scholaris* R.Br)

Family: Apocyanaceae

Vernacular name

Bengali: *Chattin*; Bengali: *Satvin*; Hindi :*Chatiyan*; Kannada: *Maddala*; Sanskrit: *Saptaparna*; Tamil: *Pala*; Telugu: *Palasi, Aedakularite Chettu, Palagaruda* ; Trade: *Chatiyan, Shaitan Wood, Devils tree, Dita Bark*

Folk Belief

The village people believe that the tree is a den of devils and try to cut it, if it is found in the villages. That is why it is called *Saitan* or devils tree. The hilly tribes of Western Ghats do not like to sit under the tree as it invites evils. Some people do not walk even under the shade of Chatiyan.

Geographical diversity & Morphological features

The trees are evergreen and found mostly in Western Ghats where rainfall occurs throughout the year. The plant prefers moist regions and reaches to a height of 25m. The bark is rough and dark ashy in colour. The branches are well spread. The inflorescence is clustered raceme. The flowers are crowded and have a light green colour. The flowers are fragrant. The fruits are in bunches and have a length of 30-60cm, hanging in pairs. The leaves are thick green, leathery and 5cm in length. The tree is called '*Saptaparna*' because seven leaves are present in a whorl but not found in single.

Economic utility

The wood is used for packing cases, tea-boxes, writing black boards usually used in elementary schools, lamin boards, minor furniture, frames, and scabbards. It is also employed in making ply wood, veneers, match splints and pencils. The wood is suitable for paper making. Wood charcoal used for gunpowder and in fireworks.

The bark also yields a fibre used for making ropes. Ash of the plant employed to open abscesses because of its caustic nature.

Bark is considered as highly medicinal and *dita* bark is employed as bitter tonic, febrifuge, anthelmintic and galactagogue. It is also used in the form of tincture extract for chronic diarrhoea, cardiac troubles, and asthma. The bark can also be used s haemostatic. The bark contains echitamine (0.16 – 1.27%) alkaloids which are helpful in digestive troubles. The green leaves are used in the treatment of beri-beri, dropsy and congested liver.

Latex usually applied to ulcers, sores, tumours and rheumatic swellings in Ayurvedic system. Flowers yield an essential oil and the alkaloid *picrinine* which acts as depressant on the central nervous system.

3.4 Chiku (*Achras zapota* L) Syn. *Manilkara achras* L.

Family: Sapotaceae

Vernacular name

Assamese: *Sopata;* Bengali: *Sabeta;* English: *Sapodilla;* Gujarati: *Cheeku;* Hindi: *Chiku;* Kannada: *Sapota hannu;* Marati: *Chiku,;* Oriya: *Sopeta;* Punjabi: *Sitaphal;* Tamil: *Sapota*; Telugu: *Sapota*

Origin

The Chiku is thought to have originated in tropical region of Southern Mexico and it has now spread throughout tropics, cultivated widely in Africa, India, West Indies, Malaysia, Philippines and Southern Florida.

Geographical diversity

The chiku tree grows best in tropics coupled with high rainfall of 130-250 cm. It also grown in drier zone of Deccan plateau and sub-montane tracts of Indo-gangetic plains. It can tolerate heavy rains and cloudy weather. High summer temperatures above 43°C may cause dropping of blossom and fruit scorching. The soil should be well-drained. Alluvial loams, on the banks of rivers, sandy loams, red laterites of heavy rainfall area are suitable for growing chiku.

It is a shallow-rooted tree and depth of soil should be atleast 5m in montane region. Soils containing high lime are not suitable as they cause loss of Chlorophyll from leaves.

Morphological features

Chiku is a evergreen tree and attains to a height of 10-15 cm, having glossy leathery leaves clustered at the ends of young branches. The leaves are 6-7 cm long, and 4-12 cm wide and entire. Interestingly, all parts of the tree contain milky latex, especially in the bark. The flowers are borne in the leaf axils but occasionally terminally also. The fruit is about the size of a large duck egg: with a brown, rough skin. The fruit contains acrid milky juice and tannin at immature stage. The flesh is yellowish brown, soft and sweet after ripening. The pulp of fruit contains 4% sugars (Leavulose, dextrose and Saccarose) and small quantity of bitter *saponin.*

Economic utility

The fruits are edible. The bark contains 25% latex which is called as chicle. It is used for the manufacture of bubble gum and other sweets. Tanin from the bark is used for colouring ships, sails and fishing boats. Timber is durable.

Place in bioaesthatic planning

The trees can be planted in the front yard of the houses, road ways and in public parks. It is an excellent shade giving tree and can be planted around industrial sites, Collage and public institutes.

3.5 Curry leaf tree (*Murraye Koenigii,* Spreng)

Family: Rutaceae

Vernacular names

Assame: Bishahari, *narasingha.*; Bengali: *Barsanga, Kariaphulli;* Gujarati: *Goranimb;* Hindi: *Kathnim;* Kannada: *Kariberu;* Malayalm: *Kariveppilei;* Marati: *Poospala, oandla*; Oriya: *Barsan;* Sanskrit: *Surabhinimba;* Tamil: *Karivempu;* Telugu: *Karivepaku;* Uttaranchal: *Gandhela*

Geographical diversity

The tree is grown in deep fertile soils preferably well drained, light textured underground layer. It prefers moderate rainfall areas and not withstand prolonged droughts. It is a common undershrub in Himalayan forests, generally in ravines and wastelands. Often the tree is gregarious in dry beds of streams and rivers.

Morphological features

Curry leaf is a deciduous shrub with thin dark brown bark. Leaves are compound, feather like, 15 – 30cm, long, rachis terete, and pubescent. Leaflets are 10-25 arranged opposite on rachis, ovate or lanceolate, oblique at the base, retuse and irregularly crenulate. Inflorescence is terminal corymbose panicles with white flowers. Fruit is subglobose, 1 cm diameter, reguose, black when ripen. New leaves are borne in March and April and comes to flower in April and May during hot months. Fruits are formed in the winter.

Economic utility

The wood is even-grained, hard, durable, capable of being used for sickle handles and other farm implements. The fresh leaves have a feted smell, but gives agreeable flavour to curries when boil in oil. The leaf is much used as vegetable and flavourant in penisular India, Where it is known as curry-plant. Leaves yield an essential oil used as fixative for heavy type of soap perfume. Fruits edible. Roots, bark and leaves are carminative, stomachic and tonic. Leaves are used for checking vomiting and dysentery. Root juice is taken by village people relief of renal pain.

Place in bioaesthatic planning

The trees can be planted in gardens, front yards of houses, hospital premises and as hedges for its elegant looking leaves.

3.6 Drumstick or Horse radish tree (*Moringa oleifera,* lann) (Syn. *Moringa pterygosperma,* Gaertn)

Family: Moringaceae

Vernacular names

Bengali: *Subunjna;* Hindi: *Sanjna, Sainjna, Sohjna, Sondna;* Malayalam: *sakhta;* Sanskrit: *Shobanjana;* Tamil: *Morunga;* Telugu: *Mulaga*

The elegant look drumstick tree is frequently not appears in the front yards of houses because of belief that one should not see the tree in the early morning which results in several obstacles in the daily life. This false belief most possibly come from its heavy infestation of leaf eating caterpillars particularly during rainy season. Sometimes, stems and entire branches covered with caterpillars which looks awkward. Fires are burnt at the tree base to kill such enormous number of worms. The trees are planted near the cattleshed, field bunds and in gardens for its tender fruits which are almost important ingredient of every south Indian dish.

Geographical diversity

Drum stick is a native of Western part of sub-Himalayas and oudh in Gangetic belt. It is grown in both tropical and sub-tropical regions. Though it is moist loving tree but resistant to water logging. Dense natural stands can be seen along the banks of Jamuna, Song and Suswa rivers and Saharanpur Siwaliks. The tree can be propagated through seed and vegetative cutting. The young plants are hardy and fast growing.

Morphological features

Drumstick is a fairly large deciduous tree with thick corky bark. The roots are pungent. Leaves are tripinnate. Pinnae opposite and has 4-6 pairs. A hairy gland is present between each pair of pinnae and pinnulae. The leaflets obovate or elliptic membranous, light green and entire. The leaves look like fern, and divided. The tree shed its leaves for a short time before flowering from February to April. Flowers comes out in large loose clusters, appears along the leaves. The flowers are creamy white and strongly honey scented. The fruit is pod-like capsule, triangular and pendulous contains 3-cornered seeds, winged at the angles. The fruits appear from May to June. The pods ripen about 3 months.

Economic utility

The tender fruits are eaten as vegetable and some times pickled. In South India, the fruit is main ingredient of *Sambar, rasam* and other juicy dishes. The leaves and flowers are also eaten as vegetables. Some wild trees produce bitter flowers and fruits, which are used in country medicine. The leaves are lopped for fodder, espicially for camels. *Benoil* is extracted from seed which is edible.

Young roots serve as a substitute for horse radish and used in native medicine. A gum similar to *Tragacanty* obtained by making incisions in the bark.

Place in bioaesthatic planning

The much divided leaves gives graceful appearance. The tree stands pollarding very well and gives excellent shade and can be planted along the farm roads, corners of streets and in public gardens.

3.7 Gliricidia or Spotted gliricidia (*Gliricidia maculata* H.B & K)

Family: Papilionaceae or Fabaceae

Vernacular names

Tamil: *Kona maram*; Telugu: *Madri, madura*

Geographical diversity

The original home of the plant is said to be tropical South America. The plant was introduced into India by British traders. It is very common border plant in South India gardens especially in Andhra Pradesh, Tamilnadu, Kerala, Karnataka and Maharashtra where warm temperatures exists most of the year. It can not tolerate very low temperatures and killed by frost which occurs in Northern India. The latin name *Gliricidia* means *Killer of rats* or destroyer of rodents as its seeds contain rat poison. The species *maculata* means "spotted" as the underside of leaves contain glandular spots.

Morphological features

Gliricidia is small, charming tree with long feathery compound leaves. The leaves are shiny, attractive and odd-pinnate. The grey branches sweep out from the base and bend downwards in arch-like manner. Black spots are present on the underside of the leaves. If regularly prunned, the tree is a best hedge plant. The fruit is long pod and flattend. The leaves fall at the end of winter (February) and starts flowering immediately. The pale pink flowers on the branches are highly charming during the beginning of march. The plant can be propagated mainly through stem cuttings and seed. The branches are highly brittle and it not cutted every year, they broken during strong winds and cyclones. Hence the tree requires frequent pollarding.

Economic utility

The tree is the best hedge plant if frequently pruned. The striking beautiful flowers add glamour to the farm houses and around industrial sites. The green leave and flowers are rich in nitrogen (3.4%) and proved to be an excellent green manure to the rice crop. The seed containing rat poison which is used in rat baits in rural areas. The cutted branches and stubbles are used as fuel. Wood is used for house posts, stakes, fences and railway cross-ties. Flowers are used as vegetables by tribals.

Place in bioaesthatic planning

Gliricidia is an excellent hedge plant around the farms and factories. The plant can be grown on either side of farm roads and canal bunds.

3.8 Jacaranda (*Jacaranda acutifolia* juss)

Family: Bignoniaceae

Vernacular names

Bengali: *Nili gul mohar*; Brazil: *Minmosaefolia*; Hindi: *Jacaranda*; Telugu: *Neeli turai*

Geographical diversity

Jacaranda is a small beautiful tree and originated in Brazil. It is distributed throughout India except North-west. Profusely flowered Jacaranda trees can be seen in some of the cities like Lucknow, Chandigarh, New Delhi, Dehradun, Allahabad, Agra and Calcutta. It grows well in well drained soils and can not tolerate water logging. Jacaranda can be grown even in high altitudes (2000m).

Morphological features

Jacaranda is a fern-like tree. It attains to moderate height, having fern-like bipinnate leaves. The leaves are similar to *Albizzia* and *Acacia* foliage. Each leaf is having several segments. The foliage is highly graceful and appreciated by all nature lovers. The tree sheds its leaves before flowering and new leaves arise after completion of flowering. The inflorescence is loose pyramidal terminal panicle. The panicle consists of lovely blue-mauve flowers of about 50-100 number with long tubes. Flowering starts in March and continue upto middle of May. The mauve coloured flowers please the viewers during hot summers. Fruit is a flat dehiscent capsule appears green but turned to brown after one year. The capsule contains several seeds equipped with papery wings for dispersal. The stem wood is fine textured with moderate hardiness. The tree is quick growing but has short life span of about 20 years.

Economic utility

The bark and leaves are used in curing syphilis and blennorrhagia. Green leaves are also used as vulnerary and infusion is employed as a pectoral. The beautiful fragrant wood is used for making knife handles, sickle handles, churning sticks, tea poys, toys and household furniture.

Place in bioaesthatic planning

The tree is very popular for its graceful foliage and beautiful flowers and can be planted in public parks, side walks of roads in towns and metropolitan cities. It can also be grown in hospital compounds, Dak bungalows and as avenues in Universities and corporate hospitals. It not only gives shade but also gives pleasant blue look to surroundings during hot summers. The beauty of tree is described in a following manner by one American traveller.

> ***"India is an unique experience and no where is available in the universe. The landscapes are beautified by densest forests and a variety of graceful flowering trees. The Jacaranda with loose clusters of deep blue mauve flowers and fragrant flowering of mangoes and neems always remembers the trees of India throughout the life."***

3.9 Mast tree (*Polyalthia longifolia* (Soln) Thw

Family: Annonaceae

Vernacular names

Bengali: *Ashok, Debdarn*; English: *Indian Fir*; Hindi: *Asopalav, Ashok*; Kannada: *Cemetery tree* ; Tamil: *Nettilingam*; Telugu: *Ashoka, Chettu*; Telugu: *Naramamidi*

Geographical diversity

The original home of mast tree is thought to be peninsular India or Sri Lanka. The drop down of its leaves suggests that the native home of the tree may be in an area of heavy rainfall. The tree can easily drop down the rainfall water with droopy leaves and smooth waxy surface of foliage. The botanical name is derived from Greek word *"Polyalthiano"* means many diseases to heal and *longifolia* refers long-leaves. The tree is often may cause some confusion with another sacred tree *Saraca indica* which is a real, Ashoka mentioned in *Ramayana*. The mast grows well in deep fertile and high rainfall areas. At initial stages, it requires frequent watering but later become resistant to drought. Luxurious growth can be seen in warm tropics. As the growth of the tree is slow, the wood is sufficiently hard to stand against speedy winds and typhoons.

Morphological features

The mast tree is a tall evergreen, very elegant tree with dark green shiny foliage and reaches to a height of 16m under favourable growth conditions. The bark is smooth and stem is straight. The branches are at right angles with the stem. The leaves are thick, black green, narrow lance shaped with wavy margins. The leaves are hairless waxy smooth, and shiny. In another variety, branches are not spreadout, but hanging rather close to the main trunk and looks like pyramid. It can be compared with evergreen cypress of South Europe. The long shiny leaves makes the viewers more pleasant than any other trees.

The tree flowers in February and fruit ripening takes place in August and September. The flowers arranged in fascicles of 5-6. The sepals are triangular shaped and petals are small and greenish in colour. Fruits are green, rounded with 12mm. diameter and turned to dark purple upon maturity. The seed viability is small and should be seeded in soil immediately after ripening. The seeds are eaten by bats during night time.

Ashok is considered as sacred tree in several parts of India and Often seen in Hindu temples. The leaves are knitted like thread and hang over them at the doors of houses on festive occasions. The tall straight trunks were used for masts. The wood is light and flexible and extensively used for making drum cylinders in South India. It is also employed for making boxes and pencils, matches, scaffolding and carriage shafts. Fruits eaten in times of scarcity. Bark is medicinal and febrifuge.

Place in bioaesthatic Planning

The straight trees are very elegant look and can be planted in front of office buildings, parking places, corners of houses, road sides of parks. These can also be planted at the edges of lawns and along the pathways of colleges and Universities. On account of graceful column with downward branchlets the tree is esteemed as good avenue tree. The plants can be grown on the southern side of houses and office buildings for the protection against sun heat, speedy winds and dust. Mast is an excellent tree for border planting in industrial sites.

3.10 Peepal (*Ficus religiosa* L)

Family: Moracaee

Vernacular names

Bengali: *Ashathwa*; Gujarati: *Jari, pipro, Pipul*; Hindi: *Pipli, papal*; Kannada: *Arali;* Malayalam: *Arachu, alrayal*; Marati: *Ashvatha, Pimpala*; Sanskrit: *Ashvatha Pipala*; Tamil: *Arasu, aswatham;* Telugu: *Ravi, Bodhi*

Popular folklore

Hindus considered the tree is most sacred and worship every morning after completion of bath. According to hindu *Puranas* the tree is the form of '*Trimurthi*' (Trinity of gods) and described in the following way:

> ***"Moolato Brahma rupaya, Madyato Vishnurupine***
> ***Agrata Sivaroopaya, Vriksharajayate Namaha"***

"The *Brahma* (creator of universe) lives at the base of the stem, *Mahavishnu* (Grower/Protector of worlds) dwells at the middle, *Mahasiva* (destroyer) stands at the top of the tree, the king of tree, I pay respects to you".

Orthodoxy Hindus not cut the tree because of the belief that cutting peepal is an evil which is greater than killing the sacred cows. In villages, the peepal and neem seedlings plant together in open places and worship them daily. When they attain considerable height, marriage ceremony was organised between them. The rural folk believe that peepal is the wife and neem is the husband. According to *Bhagavadgita.*

Fig. 3.1: Mast tree (*Polyalthia longifolia*).

Among the mountains, I am the mighty Himalayas,
Among the cupids, I am the prahlada
Among the birds, I am the Garuthmantha (great eagle)
Among the trees, I am the Ashwatha, the king of trees.

Gita -Vibhuti Yoga-

According to Buddhist literature, Gautama Buddha performed his mediation under the shade of peepal tree present in Buddha Gaya attained enlightenment of soul and revealed the secret of birth and death of human life. The young Siddhardha became Buddha (enlightened soul) on the day of *Buddha Purnima* and the tree is called the *Bodhi* from the day onwards.

The Mourya emperor, Ashoka become the follower of *Bodhisatwa* and disseminated the religion in other foreign countries. At that time, Sanghamitra and Mahindra the daughter and son of Ashoka and other monks taken the sapling of *Bodhi* from Buddha Gaya to Srilanka in 288 B.C. This tree is still living in Anuradhapura and doing well after two millenia. However, the original Bodhi in Buddha Gaya is no more. The ruling dynasties of Srilanka preserved the tree through several centuries as they believe that there is a strong relationship between the life of peepal and their political power.

Geographical diversity

The peepal is naturalised throughout the Indian subcontinent. Wild forms can be seen frequently in lower slopes of Himalayas from Punjab to West Bengal. It is also found in deciduous forests of Madya Pradesh, Maharashtra, orissa, Andhra Pradesh and Karnataka. Hot climates with more than 100cm rainfall is favourable for its rank growth. It is resistant to water streess but susceptible to frost. The strong root system reaches perennial water sources of underground and met its water requirement. In several forests, it is often seen as epiphyte at an early stage and later kills the posts. Being believe as sacred, the tree is planted near temples, shrines, banks of canals and rivers and at meeting places in the villages.

Morphological features

The tree reaches much large proportions and largest among the Indian fig trees. These are also seen on old temples, forts, old banglows and walls where bird droppings are common. Byron described them as '*green ruins dwells on gray walls*' because of their destructive nature of buildings. The leaves and young branches are smooth and shiny in nature. The leaves are leathery, broadly oval, suddenly narrowed at the apex and form a long tail. The base is heart shaped, prominent nerves on the leaf with a long stalk. Leaves are pendulous and quiver even for a slight breeze. The village peasants sits under the peepal and being lulled to sleep by rustling of leaves. Figs are borne in pairs at an angle between the branch and leaf sheath. Birds are very fond of these fruits and the seeds passed though the gut undigested and helps in dissemination.

Economic utility

The bark contains tannin and the infusion used for ulcers and skin troubles. Fruits and tender buds are eaten by tribal people. Fruits have a good medicine value and acts as laxative. Latex extracted from the stem is used to fill up the cavities in hollow ornaments and also employed for preparation of birdlime for catching the birds. Bark yields a fibre and can be used for paper making. Wood is not so hard and used for making packing cases and match boxes. It can also be used for manufacturing fellow of wheels, spoons and bowls.

Place in bioaesthatic planning

The tree can be planted in open places of villages where it spreads like umbrella. Plenty of space is needed for spreading its branches. These can also be planted along with National highways, in zoological parks and bird sancturies.

3.11 Persian Lilac (*Melia azedarach* L)

Family: Meliaceae

Vernacular names

Assame: *Thamaga,*; Bengali: *Ghora nim, mahamim* ; Gujarati: *Bokam limbodo;* Hindi: *Drek, Bakain*; Kannada: *Arebevu;* Khasihills: *Dieng-jah-rasang* ; Malayalam: *Karin Vembu, Sima Veppu*; Marati: *Pejri*; Nepali – *Bakaina*; Punjabi: *Drek;* Sanskrit: *Mahanimba*; Tamil: *Malaivembu*; Telugu: *Turaka vepa*

Geographical diversity

Persian lilac is not a native of India. The trees grows wild in Baluchistan, Persia, China, Myanmar and sub-Himalayan tracts. As the name indicates, the original home of the tree is possibly *Iran*. The plant is introduced into India during the muslim invasions of medival period. The plants are well distributed in Siwalik tracts and North West India. In Punjab, the tree is common around the irrigation wells, borewells and tanks. The lilac is well spread in India during the dynasty of Sultans.

The botanical name *Melia* means *"ash"* which is derived from Greek and *azedarach*, means *"an independent tree"*. It grows luxuriously in moderate rainfall area and relatively tolerate to frost compared to Indian neem.

Morphological features

Persian lilac grows to a height of 12-15 metres, fast growing, deciduous and elegant tree. Stem covered with smooth dark grey, slightly cracked vertically in long fissures. The tree is similar to neem with an extensive crown.

The leaves are beautiful dark green, fern like appearance, compound and divided to several pinnae. The leaflets are 5-7 and fall in winter (December-January) after turned into yellow. New leaves appears in spring (March) along with flowers and looks like new bride. Flowering occurs in spring and floral bunches appears in April.

The flowers are numerous and have lavender colour and emits mild scent from them during night. The fruit is rounded drupe and ripen in winter. Four stoney seeds are present in each drupe. The fruits hang on the tree in yellow clusters upto the next flowering season. The fruits wrinkled after drying. The seeds are having natural hole through the centre and suitable for making beads.

Economic Utility

The fruits are poisonous if eaten by cattle but yields medicinal oil upon extraction. The drying oil is used for making soaps and hair oils. The stony seeds are employed in making necklaces and rosaries. The rosaries worn by *Muslim Fakirs* and *Sadhus* are usually made with these seeds. It is believed that these rosaries drive away the evil spirits.

Leaf juice is anthelmintic, diuretic and emmenagogue. The gum oozing from trunk is used in spleen enlargement, and infusion of bark in ascariasis. The roots are said to be anthelmintic and used in native medicine.

Wood is used for making ammunition boxes, museum cases, turnery, ornamental plywood and muscical intruments.

Place in bioaesthetic planning

Being very-fast grown species, attains to a height of 3-4 metres within a year. The saplings not tonched by cattle due to bitterness of plant. This elegant tree can be planted as avenues on public roads, entrance of air ports and railway stations. These avenues are a fine sight during flowering season for the city dwellers living in concrete jungles.

3.12 Sendroi (*Buddleia paniculata* wall)

Family: Buddlejaceae

Vernacular names

Hindi: Sendroi

Geographical diversity

Sendroi do not demand any particular soil and adopted to all situations. Abundantly found in mountain slopes from 1200 to 200m altitude. Luxurious growth can be observed in exposed sunny areas and resistant to drought. It's adaptability to grow under very steep and dry slopes of hills is remarkable.

Morphological features

Sendroi is an evergreen shrub with a crooked stem and grows to a height of 2-3 metres. The bark is light grey, thin, peeling-off in vertical strips in aged trees. Branches covered with rusty-tomentose. Leaves are lanceolate sometimes ovate-triangular-leaf margin dentale or sub-entire. The shrub flowers in hot summers (April-May) and fruits develop during the start of rainy season. The inflorescence is pedunculate head or irregular panicle. The flowers are pink or white, forming cluster. The flowers are fragrant and atractive. The fruit is ellipsoid capsule.

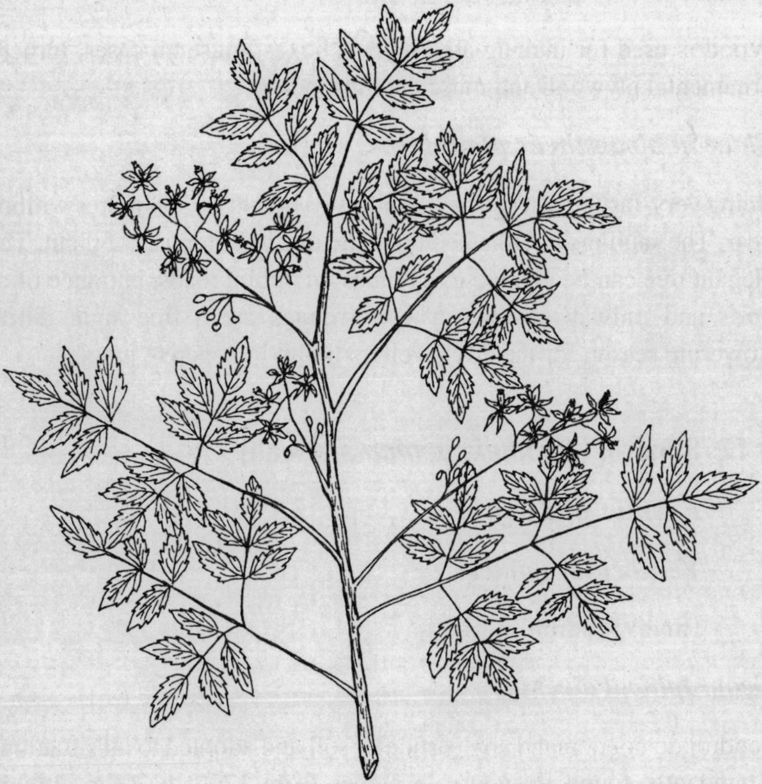

Fig. 3.2: Persian lilac (*Melia azaderach*).

Economic utility

The shoot extract is highly medicinal and used for curing ulcers of digestive canal. Leaf decoction is good remedy for dysentery and diarrhoea. It is unique species for re-stocking the bare hillsides due to its vast root growth and adaptability to water and nutrient stress conditions. The stems are used as fuelwood in country side.

Place in bioaesthatic planning

The shrub can be planted in public parks and private gardens for its fragrant flowers.

3.13 Dudleia (*Buddleia asiatica* Lour)

Family: Buddlejaceae

Vernacular name

Doon: *Mania;* Himachal: *Bhatia;* Hindi: *Dudhia*

Geographical diversity

Dudleia is very common in open dry places, besides ravines and around grasslands of Himalayan mountain systems. It is well distributed at the Himalayan siwaliks from 900-1800m altitude. The plant seldom demands particular fertile soil. It can thrive in any type of soil and well adopted even in eroded soils.

Morphological features

Dudleia is a soft, grey tomentose shrub and attains to a height of 3m. The leaves are dark green, opposite, lanceolate shaped with serrulate margins. The leaves are smooth and white tomentose present beneath the leaf. Inflorescence is dense terminal panicle with continuous spike. Flowers are small and brown. The inflorescence is attractive against the background of brilliant green of foliage. Th fruit is ellipsoid capsule. The shrub comes to flowering during hot summers. It propagates through seeds and branḉh cuttings.

Economic utility

Dudleia is highly attractive for its glossy green leaves and grown in public and private gardens as an ornamental plant. The flowers are edible and eaten by tribals at the time of scarcity. The roots are added in fermented sugar solution while preparing country liquors. The wood is tough moderately hard and employed for making farm implements. Walking sticks are made with the branches.

Place in bioaesthatic planning

The shrub can be planted in parks and public gardens on the side of pathways.

3.14 Tungla (*Rhus cotinus* Linn)

Family: Anacardiaceae

Vernacular names

Doon: *Tung*; Hindi: *Tungla*

Geographical diversity

Tungla is highly adopted in rocky and infertile soils. Highly adopted to drought conditions and variate temperatures. The shrub is fairly common in Garhwal regions of Himalayas slopes. It prefers open sunny places and found upto an altitude of 1500m.

Morphological features

Tungla is deciduous shrub and occasionally form into small tree, attains to a height of 4m. The leaves are elliptic obovate, alternate with entire margin. Leaves are glabrescent on upperside and pubescent on lower side. Inflorescence is terminal hairy panicle. The flowers are yellowish with elongated pedicels of sterile flowers. Fertile flowers are very few. Fruit is pinkish, obliquely obovate shaped, drupe, covered with hairs. The shrub flowers in hot summers and comes to fruiting during autumn.

Economic utility

The fruits are edible and eaten by forest dwellers. The leaves and flowers emit the characteristic resinous odour of mango when bruised and added in some curries. The leaves used for cattle bedding in wet areas. Cattle can be kept with leaf bed upto three days comfortably in rainy days. The leaves later composted and used as manure in field crops. The twigs used for basket and chair making.

The dark yellow coloured heart wood is employed for making picture frames, cabinets, tool handles and farm implements. The branches can be used as firewood in country hearths.

Place in bioaesthatic planning

The shrub can be planted on the wayside of schools, colleges and public parks.

3.15 Variegated Bauhinia (*Bauhinia variegate* Linn)

Family: Caesalpinioideae

Vernacular names

Bengal: *Bidul, Kovidara, Ratktakanchan*; English: *Mountain Ebony*; Hindi: *Baisakhi Kachnar*; Kannada: *Arisinantigi, Kanchavala*; Malayalam: *Kovidaiam, Chuvannamandaram* ; Marati: *Kanchan*; Oriya: *Boroda*; Tamil: *Mandarai, Seqapu Manchori*; Telugu: *Bodanta, Devakanchanam*

Geographical diversity

Variegated bauhinia is indigenous to India and grown luxuriously in drier hilly parts. It is valued as it grows in poor soils, even in open rocky areas and protect the soil from sliding and erosion. It is widely distributed in U:P, Maharashtra, Madya Pradesh, West Bengal, Andhra Pradesh, Tamilnadu and Kerala. Bauhinia is very common in Sub-Himalayan tracts (1800m altitude) from Indus towards Assam. Dense populations can be seen in Kangra valley where it is grown in every village.

Morphological features

Bauhinia is a deciduous tree grows to a height of 9m. It is an open branched and crown irregular. The bark is brown turns to rough as the age is advanced. The leaves have attractive shiny green in colour and split down in the middle. They have about 10-15 strong nerves, radiating from base of the stalk. The tree shed its leaves in the month of February. The geneled trunks clothed with rough bark gives some kind of attractiveness to the tree The tree lives about 70 years.

Flowers are big, white in colour and fragrant light yellow or pink spots or purple-mauve spots are present on the flower. Flowers are borne when the leaves completely fall-off. Fruiting occurs in rainy season. The pods are medium in size.

Economic utility

The flower buds very popular as vegetable. The buds are boiled and used in the preparation of *raita* which made with sour curd. The buds are also mixed in mutton, Chicken, Pork and prawn dishes. These are rich in protein and other minerals. The tree yields a gum, having same properties of cherry gum. Oil can be extracted from the seed.

The timber is semihard and employed for making agricultural implements. The bark is rich in tannin and used for tanning and dyeing of leather. Leaves are used wrappers for *bidis* and country cigarettes. The leaves are lopped for cattle and camel fodder.

Place in bioaesthatic planning

Bauhinia can be propagated through seed. Being fast grower, it comes to flower during second year. The trees in full flowering, present a immemorable site. It is an excellent tree for broadways, open places and public parks. If other varieties like white, mauve and yellow are mixed in avenues, they display an heavenly sight.

3.16 White Siris (*Albizzia procera*, Benth)

Family: Mimoseae

Vernacular names

Bengali: *Koroi;* Hindi: *Safed siris, Siras, Himachal Karha, Karhar, Karhai*; Kannada: *Bellate*; Malayalm: *Karuntagara*; Marati: *Kinhai;* Tamil: *Konda Vagi;* Telugu: *Tella Chinduga*

Geographical diversity

White siris prefers fertile soils with high moisture. Luxurious growth can be seen in moist localities and near the swamps.

Morphological features

White siris is very large graceful tree with a deep oval crown when growing singly in open lands. The bark is yellowish or greenish white, smooth, and pulls-off in thin flakes. Leaf-reaches 40 cm long, smooth, sometimes, pubescent with large gland near the base. Pinnae has 3-5 pairs of leaflets. The leaflets are obliquely-ovate, smooth with prominent lateral nerves. Inflorescence is 1.5 cm diameter floral head, arranged in large lax, branched corymbosely. Flowers are sessile and yellowish white in colour. Pod consists of 8-12 seeded with darker blotches. The tree flowers during rainy season (June to August) and fruits appear in winter season.

Economic utility

White siris has large white sapwood which is not durable. Heartwood is light, dark brown and takes excellent polish and highly durable. The wood is used for making rice-pounders, sugarcane crushers, cart wheels, house posts, carts, ploughs, harrows and handles of saws and knives. High quality charcoal can be made with the wood. The bark is used for tanning and mixed with milled or cereal flour in times of femine for consumption.

Place in bioaesthatic planning

The trees can be planted in double rows for the car routes which is interesting and pleasant to the travellers.

3.17 Java Champak (*Michelia Champaca* Linn)

Family: Magnoliaceae

Vernacular names

Assame: *Titasopa*; Gujarati: *Champo*; Hindi & Bengali: *Champa*; Kannada: *Sanpige*; Marati: *Pivala Champa, Sona Champa*; Nepali: *Oulichamp*; Oriya: *Chompa*; Sanskrit: *Champaka*, ; Tamil: Shembyga; Telugu: *Champakamu*; Tradi: *Champak*

Geographical diversity

Champak is a native of Java forest and grows luxuriously in damp climates. The ideal habitats consists of 2200-5000 mm annual rainfall with 80% atmosphric humidity. The tree is wide spread in tropical Asia and is very common in Western Ghats of Kerala and Karnataka and eastern Sub-Himalayan zone of Assam and West Bengal. It rarely found in deep valleys which have perennial springs. The tree is capable of growing on a variety of geological formations like quartzites, Schists, slates and gnesis. Deep moist soils with well drained sub-surface soil is ideal for its growth. The trees are very sensitive to ground fires.

Morphological features

Champak is evergreen, handsome and grows to height of 20mm with a girth of 4m. In its natural habitat, it grows to a height of 35m and surpasses all their associates with bark free and straight bole. Bark is ashy grey and turns reddish brown later. Leaves are lanceolate, glabrous and accuminate at the end. Old leaves become yellow. Inflorscence is axillary and occupied with deep yellow solitary flower. Flowers are 5 cm in diameter with strong fragrance. Fruit is capsule, 8-15 cm long with dark brown in colour. Fruit contains 1-12 brown, Smooth seeds.

Economic utility

Champak is an excellet wood for Joinery, ship and boat building, door panels and carriages. The wood is used for making musical drums in South India. One kilo of Champak attar is extracted from one hundred kilos of flowers through distilling process and used in *Agarbathi* industry. Champak attar is very famous and mixed in Jasmini scents in perfume industry. An aromatic *Sampangi* oil is extracted from Champa flowers.

Fatty oil extracted from seed is used as lubricant and illuminant. Yellow dye is extracted from flowers which is used for colouring the food stuffs. Flowers are medicinal and used as stomachic, diuretic and carminative in nature medicinal system. The stem bark is febrifuge and stimulant. Dried roots are purgative and even emmenagogae.

Place in bioaesthetic planning

The trees are specially grown in the temple surroundings for its perfumed flowers. The trees can be planted along the side of national high ways, city roads, surroundings of railway and bus stations. The trees are excellent for planting in public parks because of its scented flowers. The entire atmosphere filled with sweet scent during flowering period and gives Joy to the park visitors.

Chapter

4

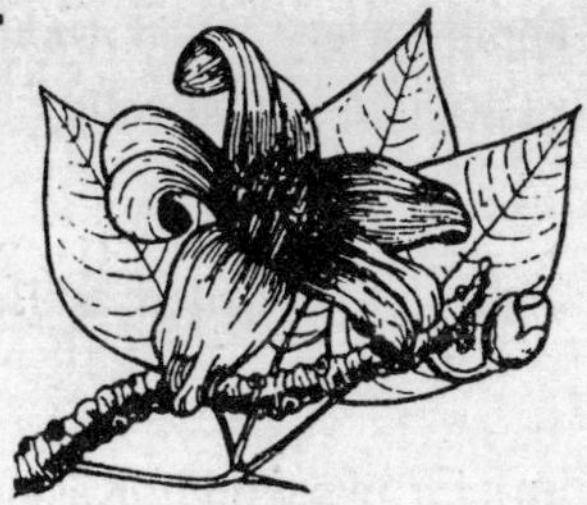

Fruit Bearing Trees

"The tree that gives luscious and flavoured fruits
is really divine gift of God
The hungry roots suck the sweet flows
From earth's breast
It is conspicuous love conveyed by nature"

4.1 Apple (*Malus pumila* Mill)

Family: Rosaceae

Vernacular names

English: *Cultivated apple*; Hindi & Bengali: *Seb, Ser*; Kannada: *Sebu, Sevu*; Ladakh: *Kushu*; Punjabi: *Seo*; Telugu: *Apple Kaya*

The apple tree originated in the caucasus mountains of South-Western Asia and has been cultivated from the immemorial. The wild forms are still exist today. The crab apple (*Pyrus baccata*) grows wild in the

Himalayan mountains. A desert apple variety called *Ambri* is said to be indigenous to Kashmir and is most popular in that valley. Three species are cultivated in India.

Malus baccata (Siberian Crab apple): It is called as Ban apple and Siberian origin.

M. baccata var. *himalaica* (Makim): Grows wild in Kashmir, Kumaun, Bhutan, Khasi and Jaintia hills.

M. sikkimensis (Hook.f) Found in Sikkim and Bhutan.

Geographical diversity

The tree is essentially suited in regions where low winter temperatures and snowfall exists. It needs significant chilling to check the rest period and induce the opening of buds in spring. It needs an annual rainfall about 75cm. The tree thrives best in well drained medium loam. It can also be grown on slopes, hill sides where soil layer is thick with impervious rocky substratum. Brown reddish brown sandy loams are also suitable for its cultivation. In normal soils, the tree bears fruits in 12 years after planting and economic life varies from 25-60 years.

Morphological features

Apple tree is a low, round crowned tree, often growing upto 15m or more, with tomentose of densely pubscent young growth, leaves mostly clustered on short shoots. The leaves are oval or elliptic, 5-10 cm long, thick, veiny, mostly short-pointed. Margins bluntly serrate. White flowers are borne in clusters, 4-5 cm across, appears with the foliage on tomentose pedicels. Fruit is fleshy pome, variable in size, shape and colour. Fruit shape is globular, oblate, conic with depressions at either end.

Economic utility

The fruit contains 10-17% invert sugars and 6% malic acid and eaten directly. Some fruits are canned and frozen which are very popular in European countries. The dried apples find use in bakeries. Apple flour is used in the treatment of infant diarrhoea. Apple juice is a popular beverage used in the treatment of peptic ulcers and diarrhoea. Jams and Jellies can also be made. Apple syrup is employed in ice cream, candy, beverage and bakery products. Cider is a popular drink which can be obtained by fermentation of apple juice and contains 0.5-8% alcohol. Apple wine contains about 14% alcohol. Apple brandy can be obtained by distilling clarified cidar in pot stills. Jams and murabba are prepared from peeled fruits. Apple pomace can also be made from apple core and peel.

Place in bioaesthatic planning

The trees can be planted along the canal banks, roads, streams and water courses.

4.2 Betel – Nut palm (*Areca catechu* L)

Family: Arecaceae

Vernacular names

Bengali – *Gua*; Gujarati: *Sopari*; Hindi & Marati: *Supari*; Tamil – *Kamugu*; Telgu: *Poka*, *Vakka*

Geographical diversity

Areca is tamil word which means a cluster of nuts. The palm is known as 'Betel palm' as it widely used along with betelvine leaf. The native home of the tree is known with uncertainty. Most of the researchers believe that areca nut is indigenous palm of India. The palm is adopted to the coastal belt upto 300km from the sea. The tree requires heavy rainfall for its successful growth. Luxurious stands can be seen in Kerala, West Bengal, Assam and other north eastern states where rainfall is heavy due to monsoon activity. Heavy textured soils with high moisture retensive capacity are more suitable. The palms should be protected from strong winds during summer.

Morphological features

Areca nut palm has straight and smooth trunk and grows to a height of 30m to 15cm thick. The leaves resemble those of coconut palm with 1.5m long. The palm comes to flower after 5 years and fruits are spherical in shape. Fruits are smooth, 4 cm across, more or less similar to poultry egg size. At earlier stage, the fruits are greenish in colour, orange at advanced stage and reddish when ripen completely. Each palm bear about 300 fruits per year. The inflorescence is cluster, born behind the leaves. The fruit is fibrous and hard nut is present inside. The collection of fruits is highly interesting, the fruit harvester climb to the top of one tree pick up its fruits and then shake the tree sufficiently so that he can hold up the neighbouring palm. The harvesting man jumps from tree to tree like *Tarzan* in the movies. In this way, he can save much time and energy, while harvesting fruits.

Economic utility

The fruit is an important article of commerce as it used along with betel vine leaf. Sometimes it is mixed with tobacco for chewing. The nut comes from tall trees has greater demand because it yields a lot of tannin. These tannins are employed for manufacturing black and red inks. The nuts are charred and used as tooth powder. The leaves are used for thatching the houses and the wood as poles of small houses.

Place in bioaesthatic planning

The palms are straight, highly attractive and can be planted on either side of pathways and narrow roads of towns and entrance of offices and hospitals.

4.3 Bread fruit (*Artocarpus altilis* (Park) Fosberg)

Family: Moraceae

Vernacular names

Hindi: *Sabji kathal*; Marati: *Vilayati Phanas*; Tamil: *Seema pila*; Telugu: *Seema, Panasa, Koorapanasa*

The bread fruit is a native of Malayan Archipelago. European voyagers to the east showed some interest about the tree products available there. Among them, British realised the value of breadfruit tree and introduced into West Indies from South Sea Islands. Later, early Dutch Voyagers introduced into India through sincere efforts from its native home. However it was now confined mainly to the West Coast of Southern region. In some of Far Eastern countries, bread fruit occupies a prominent place in the dietary habits of the people. In fact, it is said to remain as staple food in the similar manner as does in European countries. Hence the British people aptly called this as *"bread fruit"*.

Geographical diversity

As the bread fruit belongs tropical plant, it needs warm humid climate and plenty of rainfall for rank growth. It thrives well from see-level to an elevation of 900m, and flourishes on the hills of Western ghats. The climatic requirements of the tree include a mean rainfall of 250cm, well distributed throughout year, temperature range of 16°C-38°C, and a humidity of less than 70 per cent. It is intolerant of extremes of climate met at high elevations.

A good drainage of the soil is essential to maintain the trees in proper condition. That is why, lateritic red loams are best suited for its cultivation. A high humus coupled with ample fertility are the requisites for best growth of the plant. Impeded drainage stunts the trees. The fruits drop prematurely in situations of water stagnation around the roots. In rocky area, Where soils are shallow, the trees grow normally at the early stages, but later a decline sets in.

Morphological features

A fully-grown bread fruit tree attains to a height of 18m. It has large, ovate leathery leaves. The tree produces separate male and female flowers on the same tree. The male flowers produce long, yellow, pendulous catkins. The female on short, thick, club-shaped prickly structures. It later develops into fruits, oval shaped, 10-15cm diameter, green at early stage but turning brown and later yellowish when it reaches maturity. The fruits produced on short thick stalks and consists of Whitish fibrous pulp. The seeded variety is found in the wild state which is not useful for culinary preparations.

Economic utility

The fruit is rich in carbohydrate (30-98%) which is higher than that of Banana and potatos. It is rich source of calcium and is a fair source of vitamins A and B complex. The fruit is sliced after peeling the upper rind and baked before eating. The roasted fruit is preserved for using in off-seasons. The leaves provide good fodder to cattle. The tree oozes juice which contain rubber. Latex used for painting canoes and canling boats. Bark yields a fibre used for ropes, which can withstand prolonged water contact. The wood used for beams, posts, rafters and flooring of houses.

Place in bioaesthatic planning

The tree can be planted along the road sides, public gardens, water ponds and banks of canals. The tree is very famous for its thick shade during hot summers.

4.4 Coconut Palm (*Cocos nucifera* L)

Vernacular names

Hindi: *Nariyale*; Telugu: *Kobbari, Tenkaya*

Geographical diversity

People think that original home of coconut palm is India because it was mentioned in ancient sanskrit literature. In fact, its wide cultivation, use and role in religious functions made the palm as one of the representative trees of Indian subcontinent. Some authorities believe that coconut palm is indigenous to South and central America. The floating fruits might have been carried all over the pacific by ocean currents to different continents. The spreading might have taken place several centuries back and it was intermingled in the religion, cultural and traditions of pacific coast inhabitants. Coconut need enormous moisture and does well in all types of soils from sandy belts to saline patches. Sea breezes encourages the growth and fruiting of these palms.

Festivals and rituals

Arab sailors and merchants called the coconut palm as Indian nut. Marco polo, a well known foreign voyager, landed on Malabar coast in the 13th century described the tree as elephant nut because it bears fruits similar the size of human head. In good olden days, the South Indian rulers like *Guptas* and *Cholas* were active seafarers and they used the Bay of Bengal as swimming pool. They built the ships and conquerred the foreign lands and carried out business with several southeast Asian countries. They are responsible for distribution of coconuts over a wide area along with Indian culture.

The coconut festival is held at the end of rainy season in Kerala coast. The fishermen and other shore inhatitants, offer coconuts to the ocean with devotion. They worship *Varuna*, the vedic god of heavens and oceans by pouring milk, flowers and coconuts into the ocean. The rainy god, *Varuna*, represented by a pot of water with green coconut placed over its mouth.

Morphological features

Coconut palm is a tall, unbranched, straight tree, with leaf scars on the trunk. The thickness of trunk is constant, 50-75cm thick from top to bottom. The leaves are feather like appearance, 3-4cm long, with numerous leaflets on a stout axis. The leaflets are leathery, shiny green and consists of about 100 in number. Inflorescence is boat-like spathe. Female flowers are 2 times bigger than male flowers. Fruits are three sided, 25mm across and long. The outermost part of the fruit is shiny green. But some varieties have a bright orange, bright lemon yellow and reddish colours. The outer fibrous layer is water proof, very hard and beneath this hard stone is present. In side the stone, a milky-white flesh, filled with watery fluid is present.

Economic Utility

Coconut palm is a scarce tree in nature that yields many products having manifold uses. In view of this, the tree is called '*Kalpavriksha*' that means wish yielding tree. The tree yields drinking water, milk, honey, sweet sugar, oil, wine and vinegar. The leaves are used for thatching the cottages and huts to keep out the sun and rain. The village blacksmith gets his charcoal from the used nuts. The coconut water clears the wrinkles and imperfections of the face and impart rosy tints of cheeks when taken daily. The coir fibre is employed for making cordage, sofa sets, coir beds and dolls. Brooms are made from leaf ribs.

Place in bioaesthatic planning

From the decorative point of view, it is most elegant tree when planted in line on the sides of roads, around the farm houses, rest houses and public parks. The orange or yellow coloured fruits trees are highly beautiful beyond measure when planted in the middle of lawns.

4.5 Common Walnut (*Juglans regia* L)

Family: Juglandaceae

Vernacular names

English: *persian* or *European watnut*; Gujarati: *Krot*; Hindi: *Akhrot*; Kashmir: *akrut*; Telugu: *Kashmiri Badam*

The Walnut is a native of north-western Himalayas and found between the elevations of 1200-2200 m above sea-level. It also found in France, Italy, Rumania, California and China. In kulu valley of Himachal Pradesh, the planting of Walnut trees are not considered to be a auspicious. Seedlings come up in nature are allowed to grow and this has naturally resulted in the preponderance of mediocre type of trees in most of the areas.

Geographical diversity

Walnut needs a climate, free from frost in spring and extreme heats in summer. The temperature of 2-3°C or below freezing points kills the young flowers at the start of bloom. During hot summers, the nut get sun burnt and become blanks. The trees grows well in 80cm rainfall. The suitable soil for walnut is a well drained, deep silt loam containing abundant humus. Soil should not have hard pan, sandy sub-soil and alkali as the roots penetrate up to a depth of 3m.

Like pican and filberts, walnuts are monoecious i.e. the male and female flowers are borne on different branches on the same tree. The catkins (male flowers) are elongated structures which develop on the previous seasons growth. The pistillate flowers are small, borne on the tips of the first spring flush of growth of new shoots. Young trees just coming often do not produce enough catkins to supply pollen for all the pistillate flowers.

Morphological features

The Walnut is tall, handsome tree and attains to a height of 2m. The leaves are alternate, imparipinnate, 20-30cm long and composed of 5-11, elliptic to oblong-lanceolate, entire leaflets. The small flowers are yellowish-green. The male flowers occur in long pendulous catkins 5-12% long Females are found in 1-3 flowered terminal catkins. The fruit is drupe with distinct exocarp and mesocarp. The stony endocarp falls out upon ripening. The dried watnut kernal have 60% oil and 15% protein. The Walnut trees growing in nature, being of seedling origin, show great variation in growth and fruit characters. The trees grow to a height of as much as 12m. The nuts are covered over with hulls which split at the time of maturity in September-October thus exposing the nuts.

Economic Utility

Mature kernels yield a fatty oil which is used in skin ointments. Kernels consumed fresh, salted and used for manufacturing pasteries and in confectionery. Young fruits pickled, a rich source of ascorbic acid and contains 2330 mg/100. Walnut oil is used for edible purpose, artists oil colours, printing inks, varnishes and soap-making. Green hulls used for dyeing and tanning. The leaves are astringent, tonic and anthelmintic.

Place in bioaesthatic planning

The trees can be planted along the side of roads, national high ways, on terrace walls, near the dwellings and banks of lakes.

4.6 Date palm (*Phoenix dactyleifera,* L)

Family: Arecaceae

Vernacular names

Hindi, Bengali, Marati, Gujarati: *Khajur;* Kannada: *Kharjura;* Malayalam: *Ittapazham*; Oriya: *Khorijjuri;* Tamil: *Perichchankay;* Telugu: *Kharjuram*

Popular Lore of Phoenix

Phoenix is the ancient Greek name of data palm mentioned by Theophrastus. The palm is native of North Africa or Arabia and extensively distributed in Arabian countries in arid and desert lands. To the early civilizations in the valley of Euphrates, it was of still greater importance, because it formed the principal diet of the native population. The sap of the tree has been a source of intoxicating drink for the Arabs, who call it the *drink of life*. A combination of cattle milk and date fruits provide an excellent balanced diet to the desert inhabitants. Date palm cultivation is very ancient, 4000 years back along the Tigris and Euphrates rivers of Pursia.

Geographical diversity

The Palm originated around persian Gulf and became naturalized in North Africa, South Spain and North India. It is found in Uttar Pradesh, Saurashtra, Kutch, Vindya Pradesh, Rajasthan and arid regions of Punjab. The date palm is very exacting in its climatic requirements.

Successful growth occurs where long summers with high day temperatures (50°C) as well as high night temperature, a mild winters without frost, and absence of rains during flowering and fruit setting. It remarkably tolerant regarding its soil requirements. It performs well on a wide range of soils from light loam to heavy clay. Soils devoid of proper drainage may sustain plants life but growth was slow. It is very tolerant of alkali soils containing 3-4% white alkali.

Morphological features

The palm reaches to a height of 30 m making nearly a straight and shaggy trunk. It bear fruits upto two centuries. The lower trunk is marked with leaf scars. Leaves are terminal crown, arching and unequally pinnate. Being diocious, the male and female plants are separate. To have a good bearing that proper pollination of female flowers is ensured. It is completed by insects or winds and hand pollination is also required. Flowering occurs early in the spring and produce about 8-20 clusters. Each female produces 3 ovaries. The fruit is drupe and 6-7cm long. Dates are propagated through suckers and seed.

Economic utility

Date palm yields highly nutritious fruit containing about 80% invert sugars. The palm juice (Neera) is obtained from the trunk, drunk fresh by rural people. Some times it is fermented, distilled and make country *arak*, a king of spirit. The juice is also boiled and Jaggery is prepared which is used in country medicines. The stalks of leaves are made into ropes and threads. The leaflets are wrought into country mats and baskets. Stalks of leaves are made into various household articles and grain baskets. The leaves are employed for thatching, umbrellas, hats and fancy bags. The stem contains farinaceous matter and is used as food in times of scarcity.

Place in bioaesthatic planning

The palms can be planted along side of roads. The palms can be planted around farm houses, residential buildings, border of fields in arid regions which gives good look to surroundings.

4.7 Elephant apple or Wood apple (*Feronia lemonia* (Linn) Swingle)

Family: Rutaceae

Vernacular names

Bengali: *Katbel, Kait*; Gujarati: *Kavit, Kotha, Kothi, Kothan*; Hindi: *Bilin, Kait, Kavitha*; Kannada: *Bela*; Malayalam: *Vila, Vilatti*; Marati: *Kavatha, Kavith, Kovit* ; Sanskrit: *Kapittha*; Tamil: *Vilambalam*; Telugu: *Velaga*

Popular Legend

Elephants are very fond of wood apples. They swallow the fruits without breaking them. The digestive juices present in the alimentary tract of the animal dissolve the inner pulp and absorbed as it is, without disturbing the shape of wood apple. When the animal excretes through dung, people will thought that the fruit is not digested. If the woody rind is broken, only empty space can be seen like air filled ball.

Hindus believe that elephant God, *Lord Vigneswara* very fond of wood apples. They offer wood apples and perform puja with the same leaves to please the God on the occation of *Vinayaka Chaviti* festival. Monkeys also relish the fruits much. They break the woody rind by striking against the rocks.

Geographical diversity

Wood apple is a medium sized tree with strong, long branches. It attains to a height of 12m. Axillary thorns are present on young branches. Leaves are compound, 12cm long and aromatic leaf contains 5-7 leaflets on both sides. Leaflets are oval shaped with glandular dots. Leaves are thick Leathery and dark green. Flowers are light green to red and in clusters. Fruit is globose shaped woody fruit. The wood fruit is light violet to grey in colour. The seeds are present in the sweet-sour pulp. The fruits are almost cricket ball in size and very hard to break. Upper surface of fruit is rough. The tree flowers in the month of May and fruits appear during rainy season.

Economic utility

The unriped fruits are made into chutneys and curries along with grated coconut and curd. The fleshy pulp is highly tasty and made into *sherbats* in summers. In this *sherbat*, Jaggery is used instead of sugar. The drink is highly relishing and reduce heat of the body and improves digestion. The sweet *Kalakanda* is also prepared with pulp by mixing honey and boiled milk. The ripened fruit is also made into Jellies. The fruit pulp contains 15% sucrose 7% protein, 13mg calcium, 11mg iron/100g. It is also very rich in riboflavin and Vitamin C.

Fruit pulp is highly medicinal, tonic, antiscorbutic, and alexipharmic. Leaves are carminative, aromatic and astringent and useful in treating the diarrhoea and other digestive troubles. Feronia gum is also used in colic complaints. Trees lopped for fodder. Wood used for oil crushers, shoe lasts, pen holders, rulers, ornamental carving and naver of wheels.

Place in bioaesthatic planning

The trees are very beautiful during the time of fruiting and may be planted along the roads, public parks, and in Zoological parks. Avenues on farm roads, factory approach roads gives very good look during rainy season which is fruiting period.

4.8 Gum Lac tree (*Schleichera.* oleosa (Lour) Oken)

Family: Sapindaceae

Vernacular names

Assam: *Gosum*; Bengali: *Kusum*; Hindi: *Kusim, Kosom*; Kannada: *Sagade, Kendala*; Malayalam: *Pulochhi*; Malayalam: *Puvam*; Marati: *Kusim*; Oriya: *Kusamo, Swad*; Punjabi: *Same, Coorg – Chendala*; Tamil: *Pova, pu*; Telugu: *Plosuku, Busi*

Geographical diversity

Gum lac is well distributed in Indian subcontinent from the foot hills of Himalayas to Sri Lanka. Abundant stands can be found in Western Ghats, Central India, Assam and other north eastern states. Luxurious growth can be seen in mixed deciduous forests of Myanmar, java and Timor. The trees require warm weather with reasonable rainfall. These are very common along the foot of Siwaliks, lower Himalaya and Himalayan valleys.

Morphological features

Gum Lac is a large umbrageous, deciduous tree grows to a height of more than 15m. The compound leaves consists of 2-4 pairs of leaflets on strong rachis. Leaflets opposite, oblong elliptic, glabrescent and the terminal pair much the largest. Inflorescence is drooping raceme, Crowded near the branch ends. Some trees produce only male flowers. Fruit is globbose, abruptly tapering to a point, dry and indehiscent. Seeds of 1-2 embeded in fleshy whitish aril and having an acid taste. New leaves and flowers appear in the months of March and April and fruiting occurs in rainy season (August to October).

Economic utility

The young fruits of Gum Lac are edible and pickled. The acid pulp of the ripe fruit is delicious and eaten by rural people as a cooling refreshment in hot summers. Oil extracted from seed is edible and used as cooking medium in tribal areas. The oil is of great value in soap and perfumery industry. Sometimes the oil is used for hair dressing if Macassar oil is not available. At the time of flowering during the start of summer, swarms of honey bees visit the tree for the collection of nectare. At the end of winter the tree appears colourful because of its beautiful young leaves. Some of the honey sucking small birds also visit the tree and make it like a birds club.

The tree produce very highly prized quality of lac. That is why it is called as lac tree. The timber is hard and durable and employed in making oil and sugar mills, hand pestles and grinders. Flowers yield a dye. Bark powder is applied externally for itches, pain in the loins and back, ulcers and inflammation oil is used for massage in rheumatism.

Place in bioaesthatic planning

The tree can be planted as avenues in national highways, around the hospitals, Dak bunglows and office buildings.

4.9 Indian Gooseberry (*Emblica officinalis* Gaertn)

Family: Euphorbiaceae

Vernacular names

English: *Emblic myrobalan, Malacca tree*; Hindi: *Amlika, aonla*; Kannada: *Amalaka*; Malayalam: *Nelli*; Sanskrit: *Adiphala, dhatri*; Tamil: *Nelli*; Telgu: *Ratiusiri, amalakam*

The Indian Gooseberry has been popularly known to Indian folk because of its several medicinal uses. The *Emblica* genus has many species including shrubs, herbs while few are trees. Some of the species are often cultivated in green houses for their graceful foliage. The word *"Phyllanthus"* derived from greek words *Phyllon* means leaf and *anthos* means flower which can be understood as bearing of flowers on the leaves. The common name *amla* is derived from the persian word *amlah* and arabic word *ambalji*.

Geographical diversity

The tree is said to be indigenous to tropical Asia. It is commonly found growing wild in tropical forests of Indian sub-continent. Gooseberry not only found in plains but also grows well in hill slopes and mountain systems to an altitude of upto 1200 metres. It distributed throughout India except arid regions but very common in Myanmar.

Morphological features

The emblic is a very handsome medium sized deciduous tree but sometimes grows quite tall reaching a height of 19 metres or even a little more. Although it sheds its leaves, but seldom entirely deciduous. The tree is referred as evergreen tree by most of the botanists. The trunk is mostly resembles guava, with smooth grey bark and mottled stem. The feathery leaves are small, fine and delicate. The tree has a peculiarity of shedding its twigs along with leaves attached.

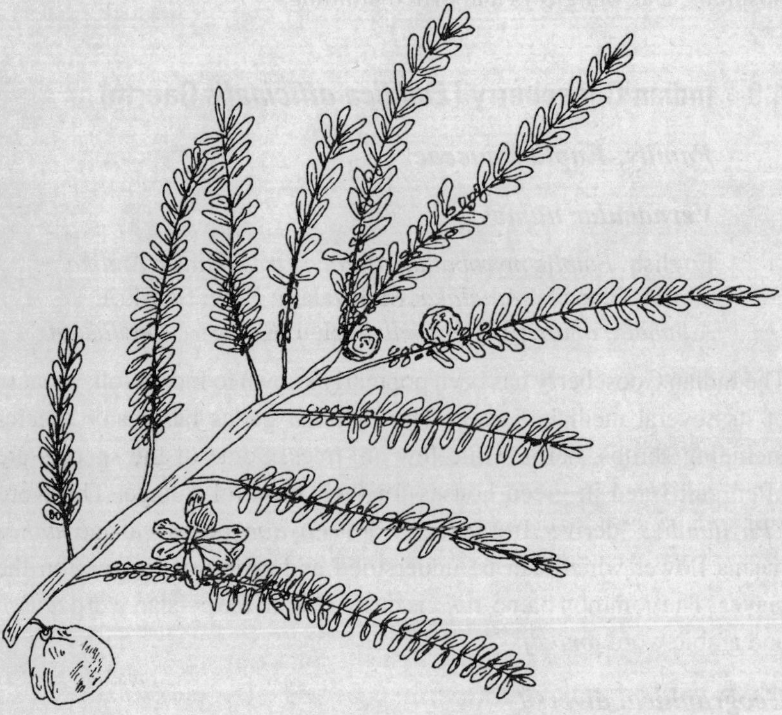

Fig. 4.1: Indian Gooseberry or Amla (*Emblica officinalis*), L – green, Fl – light green, Fr – yellowish green.

The flowers are greenish, very small and are borne in clusters often on the naked portion of the twigs below the leaves. The tree flowers from March to May and fruits ripen from November to February. The fruit is nearly stalkless smooth, yellowish, green, and fleshy. The fruit is divided into six segments through pale linear grooves. The fruit is shiny and the size varies from small seeds embeded in the flesh. Aonla is hardy plant and does not require much attention. It is a prolific bearer and Banaras variety has a handsome large sized fruit like small apple with yellowish colour.

Economic utility

The raw fruits are very rich source of vitmain C. The fruit pulp contains 1200-1800 mgm Vitamin C in 100 gms. The highly delicious *aonla murabba* is prepared in every home of South India to supplement Vitamin C. Other products like emblic sauce, candy, dried chips, Jellies, tablets are distributed to soldiers to alleviate vitamin deficiency during world wars. The fruit juice contains 320 times more vitamin C than oranges. The fruit is being used in different Ayurvedic medicines since antiquity. Fermented juice is used along with grape juice, honey and other herbs to cure various digestive and nervous diseases. The leaves are curative and used along with buttermilk, fenugreek and sour milk for curing mouth ulcers.

Fruits used in hair dyes and for shampooing hair. It also acts as detergent. Seeds yield a fixed oil which is used for making soaps. The bark and leaves are rich in tannin (21-22%). Wood used for poles and agricultural implements.

Place in bioaesthatic planning

The trees suitable for road side avenues. Planting along with national highways looks harmonious and pleasing. In villages, the tree is considered as highly sacred and used to sit and taken the dinners under the tree shade in the month of Kartika. People believe that Lord Shiva blesses boons who worshipping the tree during the aforesaid month.

4.10 *Jamun* (*Syzigium cumini* (Linn) Skeels)

Family: Myrtaceae

Vernacular names

Andamans: *Thabye*; Bengali: *Jam, Kalajam;* English: *Black Plum, Java plum, Jambolan, Jaman.*; Gujarati: *Jambu, Jamli;*

Hindi: *Jaman, Jam;* Kannada: *Nerale;* Lepcha: *Phoberkung;* Malayalam: *Perinnrel*; Marati: *Jaman, Jambul;* Mundari: *Dinkikudadaru*; Nepali: *Kalajam;* Oriya: *Jamo;* Punjabi: *Jammu;* Sanskrit: *Jamboophal;* Tamil: *Naval*, *Nerdam;* Telugu: *Neredu*

Popular Legends

The botanical name *Syzygium* is derived from greek word *Suzugos* means paired. In sankrit scriptures. Indian subcontinent was described as *Jamboo Dweepa* which means *"a land of Jambul trees"*. According to vedic puranas, Mahasiva and Vinayaka very fond of Jambool fruits. Hindus worship the Lord Sri Krishna with these fruits as it is a favourite fruit during his childhood. Some poets described, the colour of fruit is being dark like *Neel Megh Shyam*. The tree is worshipped and brahmins are fed under the shade on some festive occations as the Jambool is very dear to Dwarakadhish. Buddhists also worship the tree and plant around their monesteries. It is believed that *God Megh* (God of clouds) have been transformed into a Jamun tree.

Geographical diversity

The jamun is said to be indegenous to India and found wild in all most all parts of the country. These trees are abundant in southern parts of the country but rare in hottest areas of Deccan Plateau. From the peninsular India, the tree extends eastwords to Myanmar, Malayasia, Indonesia and Thailand and formed important component of Indo-Malaya flora. The tree is not very particular about the soil but prefers deep, well drained loamy soil. It requires hot weather for proper flowering and fruit setting. Early monsoons helps in the department and size of the fruit.

Morphological features

Jamun is a middle-sized (15m high) but attains large size in dense forests where moisture is not a limiting factor. This evergreen tree has a dark stem, shiny green leaves and dense crown. The leaves are oval to lanceolate and narrowed at the apex and base also. The translucent, pole dots appear if the leaf is seen against the sun which is a typical character of Myrtle family. The trees shed their leaves during January-February. Flowers are dirty-white, faintly scented, crowded in short racemes borne below the leaves. The tree usually blooms from March to May. Fruits are formed in june and july.

Fruit is a berry with single stone. The skin and pulp are green and highly astringent at the early stage of fruit but turns to dark purple or red. Wild fruits are large and attains to a size of 15mm x 10mm but cultivated varieties are four times smaller than wilds.

Economic utility

The tasty, sub-acid fruits are fairly nutritious and medicinal. The juice is given for diarrohoea and amoeba infections and good for all digestive troubles. The fruits also used for the manufacture of squashes, Juices, and Jellies. Seeds are powdered and given along with curd to stop the motions due to bacterial infections. The village folk make the custom that every one has to taste the jamun fruit during the rainy season as they believe that it dissolves hairs taken along with food and clean the digestive tract. The fruit juice makes the body cool and comfort by expelling the excess heat. Seeds, bark and leaves used in native medicine. Blossoms are the source of nectare for honey bees during hot summers and Jamun honey is considered as high quality medicine. The fruits and seeds are good remedy for sugar patients as it reduces the blood sugar to some extent. Green leaves are eaten by goats. Bark is used for colouring fishnets. The wood is very hard and employed for making railway sleepers, props in mines, boat building, house construction, cart wheels, toughs and agricultural implements. The green leaf is used as green manure in some parts of the world.

Place in bioaesthatic planning

The evergreen, graceful tree can be planted as avenues in University campuses, city roads, public parks and along the canal banks. It gives pleasant green look during hot summers and soothens the hearts of travellers. The dark black and violet shiny fruits with a background of green leaves is a joyful scene during rainy season.

4.11 Mango (*Mangifera indica* L)

Family: Anacardiaceae

Vernacular names

Bengali: *Arm*; Gujarati: *Amri*; Hindi: *Am*; Kannada: *Maru*; Malayalam: *Cutam, Amram, Maru*; Marati: *Amba*; Sanskrit: *Amra chuta*; Tamil: *Manga, man*; Telugu: *Mamidi, Mavi*

Few other trees have the historic background of the mango which are so intimately connected with Indian folklore. Any literature on Indian trees would not be complete if the mango tree was excluded. It is most popular, beautiful and choicest fruit tree of India and undoubtedly one of the best fruits of the world. In many villages and towns, the tree forms sitting spot where the life of community carried on like a peepal tree. Gossip of folk is exchanged under the shade of mango during afternoon hours.

Legends

Mango is highly celebrated tree in ancient sanskrit literature. Several poets referred the tree as *Wish-granting tree* and its name has been used as a symbol of love and tranquility. The season of blossoms are vividly described by famous sanskrit poet Kalidasa in his *Meghsandesh*. The flowers are considered as darts of *Manmada*, the god of love by very many writers of excellence. The fine art and sculpture turned around the tree and left marvellous representations of it. The fruits and leaves of mango can be seen in the architecture of *Sanchi stupa* built by Ashoka the great around 150 B.C. Several birds like *Koyal* starts singing after eating the mango tender buds only. The jhumming honey bees are activated by first flush of mango gardens.

Several European travellors tasted the delicious mangoes and described them as *Theobrome* meaning the *food of angels*.

Geographical diversity

Mango is a indigenous to South Asia, especially at the base of the Malayan Archipelago from where it was spread to almost all corners of the universe. The name *Mangifera indica* given to mango means "the Indian Mango bearing tree" derived from malayan word *'Mango'* or Tamil 'Mango'. The portuguese sailors spreaded the fruit into Europe from Malaya. The tree has been found wild all over peninsular India from the lower slopes of Himalayas in the North, Judging from sculpture, the tree has been cultivated in India for thousands of years. However its origin is still a matter of dispute among scientists. Mango blossoms and fruits better in regions where there is a good rainfall during june-september, followed by dry weather for remaining months. It does not thrive in the hills where elevations are more than 900 metres.

As the mango tree develops a deep and widely-spread root system, it requires a very deep but well drained soil. Some of the best mango groves are to be found on the great Gangetic plains of north India, also on the banks of large rivers.

The number of varieties grown in India is very large. Each variety has its own peculiar taste, flavour, and consistency of pulp for which it is liked by a section of populace. A desert mango should possess a pleasant, sweet, luscious, sub-acid, appetizing taste and a good flavour. The most popular varieties in India are Alphonso and Pairi in Maharashtra, Langra, Dusheri, Malda, Safeda, Samar Bahisht, Chansa and Fazli in U.P, Langra, Gulab Khas, Zardalu in Bihar, Himayuddin, Alampur Baneshan Mulge, Totapari, Rumani, Suvarnarekha, Neelum in A.P, Kesari, Jamadar, Salebhai. Amdi in Gujarat. Jehangir of A.P. Cowasji patel of Maharashtra, Mulgoa of Tamilnadu, Hathi-Jhool of U.P. are well known for their big size and good eating quality.

Morphological features

The mango is large tree and attains to a height of more than 15m. The branches spreads long distances. In densest forests, the tree grow to gigantic sizes of more than 35m. Leaves are alternately insert along the branches, but more dense near the ends. The leaves are shiny, lance-head in shape, edges are wavy and stiff in texture. The petioles are 5cm long. The inflorescence is yellowish panicle with small inconspicuous flowers. The fruit is drupe with large flesh. Large stone is present in the middle of flesh. Unripe fruit is heart shaped and possess strong favour of turpentine.

Mango can be propagated through seed and vegetatively. Propagation through inarching, budding, crown grafting and side grafting is also common.

Economic utility

It is undoubedly one of the best fruits of the world. Ripe mangoes are exceedingly refreshing to eat and is excellent source of vitamin A and C. The green fruit is used in curries, pickles and chutneys. Cotyledons used as food at the time of scarcity. Wood is used for making ceiling boards, window frames, tea-chests, brush-backs, plywood, shoe-heels, beams, rafters, window shutters. Bark is used for uterine haemorrhage and seeds are employed in asthma. The fruits are used in constipation because of its laxative property. The bark is used in Havan (sacred fire) which is one of the ritual to please the Gods of Heaven

Place in bioaesthatic planning

These trees can be planted in double avenues in big cities. These also can be grown as shade giving trees on both sides of national high ways. Pure avenues gives beautiful skyline and pleasing effect in University campuses and Industrial sites.

4.12 Dilenia or Chalta (*Dillenia indica* L)

Family: Dilleniaceae

Vernacular names

Assame: *Outenga*, *Chalita;* Bengali: *Chalta*, Marati; Gujarati: *Karambel*, *Karmal;* Hindi; Kannada: *Betta Kamigala;* Nepal:

Ramphal; Oriya: *Uva;* Sanskrit: *Bharija;* Tamil: *punna;* Tamil: *Uva*; Telugu: *Pedda kalinga, Uva*

Geographical diversity

Dillenia has a wide range of distribution in India from Sub-Himalaya tract to peninsular India. The tree grows well where the Day temperatures ranging from 38°-48°C, night temperatures 0-18°C and annual rainfall is 800-5000 mm. The trees are stunted in growth if the rainfall is less than 100 mm. The tree is very common is Bihar, Bengal and Assam and extending upto indo-China, Java and Borneo. The tree does not demand for any particular land and soil.

Morphological features

Dillenia is a medium sized, erct, evergreen tree with rounded crown with spreading branches. The foliage is bright-green and attractive. The bark is reddish, Smooth and Peeling-off in scales. The leaves are large, broad, pointed at the tip and margins are toothed which grows at the end of branches. The leaf veins, closely set and run parallel to the midrib and shows a deeply fluted surface. Flowers are large, white and fragrant and produce at the end of branches during early monsoon season (July). The fruits are hard and large which were eaten by birds greedily. Seeds are round, smooth and brown in colour and can be planted in monsoon successfully. The tree flowrishes in damp localities and frost less situation.

Economic utility

The wood is moderately hard and used for making plants, rafters, tool handles and boats. The wood is an excellent for making oars, dugouts, heavy packing cases and well-curbs. It yields excellent charcoal of high thermal energy. The Thickened sepals on the fruits used as flavouring, or made into Jams and jellies. The Juice is sweet and with sugar and taken as cool drink. Fruit is laxative and rich used vitamin C. Bark contains 10% tannin and as feed for *tusser* silkworms.

Fig. 4.2: Chalta (*Dillenia indica*), Bark –red, L – bright green, Fl – white.

Place in bioaesthatic planning

It is highly suitable for growing in the compounds of hospitals, court yards, Bus stations and compounds of Dak Bungalows and houses.

Chapter

5

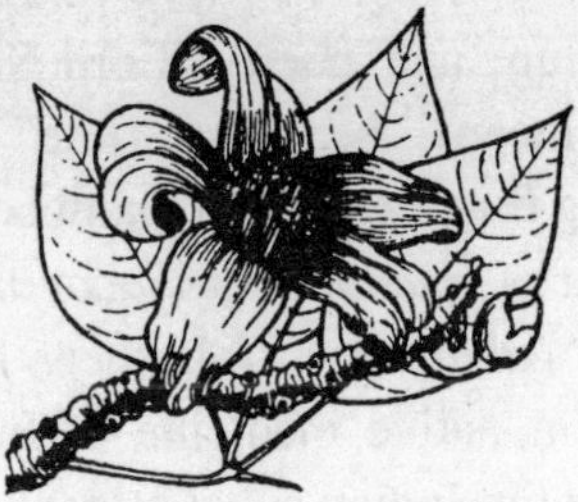

Beauty of Bamboos

"One can believe the presence of God
When melodious music flows through bamboo flute
The heavens rejoice with the divine song
The earth slept silently with sumptious feast"

5.1 Bamboo (*Bambusa* species)

Family: Poaceae

Vernacular names

Assamess: *Kotoha;* Bengali: *Ketua, Kutuasi*, Telugu: *Veduru bongu, Veduru;* Burmese: *Kyakatws;* English: *Thorny bamboo*; Hindi, Marati, Gujarati: *Bans;* Kannada: *Biduru;* Sanskrit: *Vansh;* Tamil and Malayalam: *Mungil*

Ancient Connections with literature

Though many species of bamboo are native of India, it was not found in ancient European writings before 400 BC. Bamboos have age-old connections with sanskrit literature and poetry. Lord SriKrishna is known as "*Vamsidhar*" which means "bearer of flute" because he mesmerized entire folk by playing flute, made with bamboo at the bank of Yamuna. Ancient Indians call it as '*mamboo*' or '*Sacar mamboo*' and used the plant to extract drug '*Tabasheer*'. Tabasheer has enjoyed considerable reputation in ancient native medicine of India. Nearly 141species of bamboos are present in Indian subcontinent.

Geographical diversity

Bamboos are arborescent grasses belonging to the tribe *Bambusae*. They thrive best in monsoon forests of tropics and attain maximum growth. These grasses dwindle into undershrub in cool temperate regions. At high altitudes of more than 3600m., bamboos have little growth and appears like small grasses.

In most of the southern states, Orissa, Assam, Bengal and western India, *Bambusa arundinaceae* is a dominent species. *Bambasa Vulgaris* was mostly found in some parts of Assam, Myanmar, West Indies, South America and Java. *Bambusa tulda* and *B.balcooa* are natives of Bengal. *Bambusa polymorpha* extensively occurs in Assam and East Bengal. *Dendrocalamus strictus* is most common in deciduous forests of India. In the valleys of Eastern Himalayas, luxurious growth of *Pseudostachyum polumorphum* was found. Similarly, cool loving species like *Arundinaria wightiana* and *A.aristata* are widely found in Nilgiris and eastern side of Himalayas. The major slice of bamboo species are originated in India, Myanmar, Malayan peninsula and South China and later spread into Europe and Australia.

Morphological features

There are about 30 genera and 550 species inhabiting in humid tropics. Bamboos attain to a height of 6-15m and characterized by woody pointed stems called as culms. Culm growth is very rapid (8cm per day) and in some species like *Dendrocalamus giganteus*, the growth rate is 40cm/day. The culms are round smooth and hollow in female bamboos. The male bamboos like *Dendrocalamus strictus, Arundinaria prainii,*

Oxytenanthera stocksii have solid culms. Bamboos flower only once in their life time for about 25-50 years. After flowering, they die out soon. Flowers are green and inconspicuous. The seed is indehiscent. The fruit is caryopsis and resemble like wheat grain. Bamboos are propagated through rhizomes, clump division, can layering and seed.

Economic utility

Bamboo canes are extensively used for construction of houses, cattle sheds and grain storage structures in place of timber. It is called as *"poor man's wood"* because of its extensive use in thatched houses in villages. Large bamboos are used in roofs, slads, lintels and sunshades. It is also used as reinforcement material in concrete constructions.

Chair, table and other house furniture made with canes are very famous in urban areas. Several articles like umbrellas, toys, musical instruments, bows, arrows, spears, fans, kites, tent poles, axles are made with bamboo. The pulp is main raw material for the paper industry. The pulp is suitable for the manufacture of better grade printing paper and rayon pulp.

Bamboo seeds are eaten by tribals. Tabasheer, a siliceous secretion found in the culms is used as aphrodisiac and cooling tonic. It is also used in curing asthma, cough and other debiliating diseases. Young shoots are pickled and made into curries. The leaves are fed to horses for coughs and colds.

Crooked and knotty culms used as paper pulp. Leaves chewed to relieve toothache. A paste of roots applied to boils and glandular swellings.

Place in bioaesthatic planning

The trees can be planted on the banks of water ponds, parks and waste lands.

5.2 Golden Bamboo (*Bambusa vulgaris*)

Family: Poaceae

Vernacular names

Bengali: *Basini Bans*; Marati: *Kalaka*; Tamil: *Ponmungil*; Telugu: *Pasupu Veduru*

Wonder gift a nature

Bamboos are a valuable gift of nature to man. Bamboos are fulfilling the material needs of the people since several centuries and have significant place in rural economy. They have age-old connections with artists, poets, craftsmen, muscians, and scientists and fascinating them. Surprizingly, mention has been not made about bamboo in ancient writings upto 400 B. C. Many species about 141 species of bamboo are native of India and ancient Indians called them as '*mambu*' Carcia (1563) mentioned the drug '*Tabasheer*' or *Sacar mambum*' which is a siliceous secretion from the culms in his book "*colloquois des simple drogase couses medicinaes da India*". Bamboos can be split and made into innumerable useful articles with ordinary hand tools. It is closely interwoven with the life of people in several ways and occupy a place of pride in India and other South East Asian Countries. Out of 50 principal genera, roughly twenty genera of bamboos are found in India. They are: *Arundinaria, Bambusa, Dendrocalamus, Dinochloa, Indocalamus, Melocalamus, Melocanna, Ochlandra, Neohowzeaua, oxytenanthera, phyllostachys, Pseudostachyum, Schizostachyum, Semiarundinaria, Sinobambusa, Teinoztachyum and Thamnocalamus*

Geographical diversity

Golden bamboo is pantropical origin, frequently cultivated as an ornament. It runs wild in warmer areas.

Morphological features

The culms do not bear any branches for a considerable height above the base. The culms are provided with culm-sheaths which remain attached to the nodes, till the culms become hard and then they fall-off. All species are green when fresh but this species have a beautiful golden colour with green stripes or variously variegated. *B.vulgaris* is mostly found in Assam, Myanmar, Java, South America and West Indies.

Economic utility

The young succulant shoots have for centuries been held in high esteem by Asian people as an edible delicacy. The culms are used for boats and rafts. Numerous articles like Yokes, axles, cordage, tent-poles, sticks, brushes, pipes, fans, umbrellas, kites, spears, lance shafts, bows and arrows and muscical instruments can be made *Tabasheer* or *bans lochan*, a siliceous secretion found in the culm of bamboos are used as medicine. It is used as aphrodisiac, and for curing asthma, cough and other debilitating diseases. The flowering culms exude a sweet, white brittle gum which is eaten by tribals. Bamboo seeds are eaten by tribals. Bamboo reinforecement was used in concrete piles in some countries.

Place in bioaesthatic planning

These can be planted at the periphery of lawns,

hedges and corners of Dak bungalows, industrial sites and water tanks.

5.3 Ringal small bamboo (*Arundinaria falcata* Nees)

Family: Poaceae (Gramineae)

Vernacular names

Himachal: *Go-Ningal*; Hindi: *Ringal*; Telugu: *Pipu veduru*

Geographical diversity

Ringal bamboo is more common in high altitudes, between 1500m to 3500m. in Tehri-Garhwal region of Himalayas. They are more gregarious in nature.

Morphological features

It is a small bamboo and attains to a height ranges from 2.5 – 3.5m with a culm diameter of 10-15cm. The stem is cylindrical, covered with white bloom, swollen much at nodes. Internodes are 12-30cm long. The margins, of culm sheaths are ciliate. Leaf blade is subulate, recurved with toothed ligule. The grass produce 3-4 leaves contains pellucid glands in rows between the nerves. Inflorescence is panicle borne on separate leafless stems. The spikelets present 3-4 together. It flowers with an interval of 30 years.

Economic utility

The stems are used for hookka – Pipes,

fishing rods, and for basket making.

5.4 Deo Ningal bamboo (*Arundinaria spathiflora*, Trin)

Family: Poaceae

Vernacular names

Himachal: *Ringal*; Hindi: *DeoNingal;* Telugu: *Sofa veduru*

Geographical diversity

These small bamboos are very common at higher altitudes of 3000-4000m. They are eminently gregarious in habit. It flowers for every 10 years but gregariously.

Morphological features

Do Ningal grows to a height of 4-5m with a diameter of 2.5cm. The culm is smooth, glaucous green when young. The nodes are prominent. The sheath of culm is erect with fimbriate ligule. The margins are scabrous. Spikelets 2.5 – 6cm long, pedicelled, 2-3 together in a narrow spathae "*spathiflora.*"

Economic utility

These bamboos are used for making baskets, winnows, chairs and tables.

5.5 Ningal bamboo (*Arundinaria Jaunsarensis,* Gamble)

Family: Poaceae

Vernacular names

Himachal: *Ringal*; Hindi: *Ningal;* Telugu: *Galam Vedurn*

Geographical diversity

The Ningal bamboo can be found at higher elevations ranging from 3000-3500m, Where drought prevails during most of the year.

Morphological features

Ningal bamboo is very handsome little single culm plant. The bamboos arise from a long creeping jointed rhizome. The length of culm is 5-6m, with 1.5cm diameter. The nodes are marked by a narrow ring. The sheath of culm is glabrous, yellowish in colour and narrowed to a truncate. Two bristly auricles are present at the tip. The leaves are lanceolate, acuminate, and has numerous transverse veins between the nerves. Flowering not seen yet.

Economic utility

The bamboo are used as fishing rods mats and making baskets. Split culms are employed for building cottages.

5.6 Bans Bamboo (*Bambusa nutans,* wall ex Munro)

Family: Poaceae

Vernacular names

Hindi: *Bans*

Geographical diversity

The bans is extensively seen in low valleys of Himalayas and cultivated in plains.

Morphological description

These are large bamboos, usually growing in clumps. The culms raises from a creeping rhizome but not crowded. The stem grows to a height of 8-15. with 3-8cm diameter. The culms are straight with graceful nodding covered with hairs. The stem is thick walled and unshiny. The sheath of culm is rounded at the top with black hairs on the back. Leaf blade size varies with the position on stem.

Margins are recurved. Two bristly auricles are present. Leaves 6-12, with a scabrid tip. Inflorescence is a stiff panicle which bear interupted clusters of spikelets. Many spikelete are empty with buds inside. Grains are oblong with hair on the top. It flowers periodically at a long intervals.

Economic utility

The bamboo is cultivated in gardens for ornament. It also grown in sacred groves and village wastes. Being straight and thick walled, it is highly useful for use as rafters and shafts of wooden houses. It is also employed for making dog carts, *ekkas* and cattle feed trough. The creeping bamboo used for the production of mechanical pulp.

5.7 Kaghzi Bamboo (*Dendrocalamus hamiltonii Nees* & Arm)

Family: Poaceae

Vernacular names

Assame: *Kokua*; Bengali: *Pecha*; Bhutan: *Pasting*; Hindi: *Kaghzi, Bans*; Lepcha: *Pao*; Nepali: *Tama*

Geographical diversity

Kaghzi bamboo mostly found in the forests of lower hills where rainfall is moderate. It flowers both annually and gregariously.

Morphological features

Kaghzi bamboo is a large and ugly looking species with culms slanting on all sides, sometimes curved downwards. The bamboo grows to a height of 35 m with a culm diameter of 12-15cm. Long stout branches are produced at upper parts. The culms have white-pubescent when young and turns to dull-green when old. Root scars are marked at nodes. The stem-sheath is glabrous, with patches of adpressed hairs with acute auricles. Sheath blades are lanceolate within curved margins. Leaves broadly lanceolate, smooth above, rough beneath, minutely serrulate at the edges. Inflorescence is very large panicle with branches furrowed on one side. Spikelets are ovoid, purple, depressed with 2-4 fertile flowers. Grain is ovoid and beaked.

Economic utility

The bamboo is not suitable for building purposes because of its thin walled and soft in nature. It is much used for mat-work, basket and furniture making. The young shoots are eaten as vegetable and prepare soup. The garo tribals cook rice in hollow internodes which they place over fire in a slanting position. After the charring of bamboo, the bamboo was split open to taken out cooked rice. Because of long internodes and large lumen, it is suitable for water conduits.

5.8 Kanta Bamboo or Thorny bamboo (*Bambusa, arundinaceae*) wild

Family: Poaceae

Vernacular names

Assame: *Kotoha*; Bengali: *Ketua, Kutuasi,*; Hindi: Kanta Bans; Kannada: Biduru; Marati, Gujarati: *Bans;* Sanskrit: Vansh; Telugu: Bongu veduru, Pante-veduru

Geographical diversity

The Kanta bamboo can be grown in almost all well drained soils of plains and hilly areas.

Morphological features

It is a tall thorny bamboo, grows to a height of 30-40m. With a diameter of 18-22cm. The culms are crowned and rasies from branching rootstocks. The culms are varying in size with bright shiny green. The nodes are prominent. Lower nodes are almost leafless spinescent with zigzag horizontal branches. The culm is thickwalled with a small cavity. Depression present near the base of branches. Culm sheath together with the blade is triangular in shape. The blade is covered with golden hairs when young. The margins are wavy. Leaves are small, linear or linear-lanceolate, glabrous above, glabrate beneath with scabrous margins. Leaf tip is sharp and stiff. Inflorescence is an enormous panicle with cluster of spikelets. Grain ending in a short beak. Flowering occurs gregariously at an intervals of 30 years.

Economic utility

Kanta bamboo, being superior in size and has great stength, eminently fit for all building purposes. Owing to thorns and crowded conditions of clumps it is difficult to pull the bamboos out except in small pieces. The crooked and knotty culms render it as second class bamboo and used mainly for floating heavy timber. Young shoots are pickled or made into curries. Leaves and twigs are used as cattle fodder. Leaves given to horses for coughs and colds. Grains are consumed during the time of scarcity.

5.9 Male Bamboo or solid bamboo (*Dendrocalamus strictus,* Nees)

Family: Poaceae

Vernacular names

Bengali: *Karail;* Bihar: *Banskhurd, Narbans;* Gujarati: Nakorvans; Hindi: Bans, Kaban; Kannada: *Kiribiduru;* Malayalam: *Kal*; Marati: Bharivel; Sanskrit: Vansha,; Tamil: *Kalmungil;* Telugu: Sadanapa veduru; Trade: Manwel

Geographical diversity

The male bamboos are indigenous to Siwaliks and found abundantly in Haridwar region. It is the only bamboo species found in North-West India which can tolerate to water stress during droughts. Few clumps of here and there flower every year in April.

Morphological features

It is medium sized densely tufted, gregarious bamboo. The height varies with climate and rainfall and ranging from 8 to 16m. and diameter of stem varies from 2.5cm to 10cm. The culm is darkgreen, covered with white waxy bloom when young and turned to yellowish later. The culm is characteristically crooked and twisted in congested clumps. Nodes are swollen and roots are borne at the lower nodes. Internodes are solid with small capacity. The sheath of the culm is paraboloidal, glabrous with yellowish-brown hairs. Leaf blade is triangular and covered with small hairs. Leaves are linear-lanceolate, rough, hairy on both surfaces with prominent midrib. Inflorescence is large panicle with cluster of spikelets. Many are the sterile Flowers. Spikelets spinescent, lanceolate, green and not depressed. The grain is ovaid.

Economic utility

Male bamboo is used for a variety of purposes. It is very strong and durable and would easily be considered as best bamboo but for its short size. It is supplied to the basket making and furniture manufacture industries. Being solid, it is used for making lathis, lance shafts, axe handles, industrial implements, walking sticks, mats, post rafters in temporary buildings. The pulp is used to make superior Quality of writing paper. It can also be used as umbrella handles and water bottles. The leaves serve as good fodder for elephants. Fishermen use the bamboo to build rafts and masts of country boats. It is used as a buoyage of heavy timber in rafting. The stem is used for preparation of active carbon. Pulp is highly suitable for rayon industry.

Chapter

6

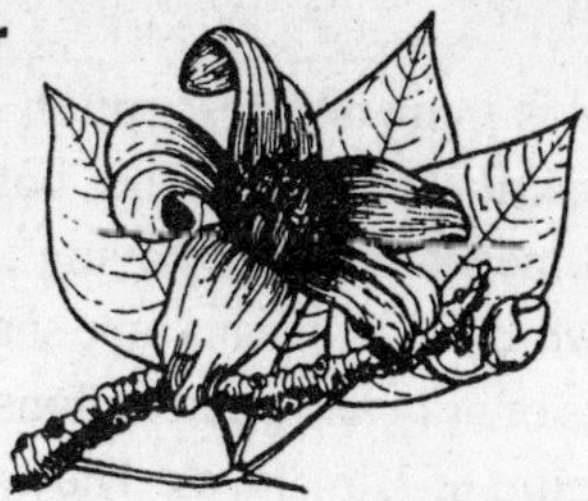

Trees of Beautiful Blooms

"Loveliest of trees, the amaltas
A tree that may in hot summer wear
The hanging blooms along with boughs
A splash of bright yellows from leafless tree
Akin to wearing golden ornaments by
handsome half naked lady"

6.1 Amaltas, Indian Laburnum (*Cassia fistula* L)

Family: Caesalpiniaceae

Vernacular names

Bengali: *Sundali, amultas*; Gujarati: *Garmala*; Kannada: *Kakke*; Marati: *Bahava*; Sanskrit: *Suvarnaka, rajataru* ; Hindi: *Amaltas, cirimalah*; Tamil: *Konnei*; Telugu: *Rela*

Geographical diversity

The original home of amaltas is India and description of its flowers are found in ancient sanskrit literature. The botanical name "*Cassia*" is greek name which means "the tree having aromatic bark like wild cinnamon". The latin word *fistula* means pipe shaped pod. The tree is found in deciduous forests of hilly and plain regions. Amaltas are also common in the Siwalik region of Himalayas. These are spread in neighbouring countries like Myanmar, Thailand, Vietnam, Java and Philippines.

Morphological features

It is a medium sized, open branched deciduous tree with glabrous, greenish-grey bark. The bark is smooth upto middle age but turn to rough and brown in old age. The leaves are large, paripinnate, 0.25 – 0.5m long with terete rachis. Each leaf is having 4-8 pairs of ovate leaflets. The new leaves are having copper colour on upper surface and a soft downy undersurface. The young leaves are pendulous and folded until the full growth. It shed its leaves in March-April months. Inflorescence is simple raceme. Conspicuous flowering can be seen during hot summers particularly in April and May. At the time of flowering the tree is completely nude by shedding its leaves. First splash of bright-yellow or golden yellow can be observed immediately after the shedding of last leaf. The flowering continues till the fresh leaves are born one can not forget the scene of flowering when it is planted near the lakes and ponds as the blooms reflect in the water.

Pods are 0.5-0.75m long, 2.5cm diameter cylindrical, smooth hard and indehiscent These are turned to dark-brown later. Seeds are compressed albuminous and parallel with septa.

Economic utility

The heart wood is yellowish or brick red, very hard and suitable for making carts, agricultures implements, posts and beams of buildings. The dark brown sweetish pulp of the ripe fruit is a strong purgative. The gum exuded from the stem is astringent. The bark is used in tanning the hides and dyeing of leather and jute fibres. The fruit pulp is used for flavouring tobacco. The *santal tribals* found in Bihar forests used the flowers as food and vegetable. The twigs are often lopped for cattle fodder.

Place in bioaesthatic planning

Amaltas is very handsome tree with thick green leaves and bright yellow flowers. These can be planted around factories, hospitals, car parking places, zoological gardens, along the roads, canal banks, railway platforms, approach roads of aerodromes and in National parks.

The germination of seeds is poor but boiling the pods in hot water soften the hard seed coat and enhance the germination. Being a xerophytic it can withstand drought and cannot be eaten by cattle, sheep or goats. The young trees comes to bloom after five years of planting. Fruits formed during cold winters.

6.2 Burmese Cassia (*Cassia nodosa* L)

Family: Caesalpiniaceae

Vernacular names

Hindi: *Burmese cassia*; Telugu: *Gulabi cassia*

Geographical diversity

Burmese cassia is indigenous to Myanmar and Malayan penisula. The tree is extensively found in those countries. It is also common in Northern India, West Bengal and Assam. The tree is slightly shorter than Java cassia. The tree prefers well aerated deep fertile soils with normal rainfall.

Morphological features

Burmese cassia is a large tree with spreading crown. The trunk is short and gnarled. The compound leaf is pinnate and about half a meter in length. The leaf comprised of 6 to 14 even pairs of hairless leaflets which are lance shaped. The odd terminal leaflet is absent. The tree comes to flower during hot summers. Before flowering the tree sheds its leaves completely.

The flowers are pink and borne in large clusters from the nodes. These clusters looking like bunches of roses and covers the entire branching system. The long stamens are swollen in the middle. It is a most beautiful tree among all the flowering plants. It flowers during june. The seeds should be rubbed on a stone till white colour comes before seeding. Otherwise it fall to germinate upto few months.

Economic utility

Burmese cassia wood can be used for making agricultural implements and handles of sickles and saws. Seeds used for external application of some skin troubles. Seeds contain a fatty oil which can be used as lubricant in country carts.

Place in bioaesthatic planning

The tree can be planted as avenues in cities and towns. It presents a unique spectacle of beauty in pink blossoms during hot months. These are highly suitable for public parks, picnic spots and medium sized gardens. Beautiful avenues can be seen in chandigarh and other North Indian towns.

6.3 Climbing Bauhinia (*Bauhinia vahliiut* & Arn)

Family: Caesalpinieae

Vernacular names

Bengali: *Sihar;* Hindi: *Maljhan, malo*; Marati: *Chambul;* Telugu: *Adda*

Fig. 16:
Dillenia aurea

Fig. 17:
Amaltas or Indian Laburnum
(*Casia fistula*)

Geographical diversity

Climbing bauhinia is very common in Himalayan and western Ghats forests. The climber prefers moist regions coupled with light shade. It grows upto an altitude of 1200m. in Himalayas. These climbers can be seen in dense forests of Western Ghats, Southern slopes of Himalayas and Eastern Ghats. The climber often attains an enormous size and stem reaches to a girth of 2-2.5m.

Morphological features

Climbing bauhinia is a gigantic evergreen climber going over the tops of highest trees in the forest and creates shade to the host tree. It is a very elegant plant, But suffers persecution from tirbals who collects the seed and consume as food. The branchlets terminates in a pair of revolute tendrils. Young branches, petioles and underside of leaves are grey-velvety. Leaves are varying in size, 7cm across, as broad as long, deeply cordate, 11-15 nerved, dark green and glabrescent. Leaf lobes are obtuse and rounded. Flowers come out in large bunches towards the end of branches. The flowers are white and attractive. The entire plant is very showy on account of its hooked tendrils and massed flowers. Pods are flat, woody, 25cm long, 10cm wide, rusty velvety and contains 8-12 seeds. The stem consists of a mass of phloem tissue which is traversed by stands of Xylem of various thickness round the pith. The climber flowers from April to June and fruits are formed in cold season. Foresters usually exterminate the plant for the protection of woody trees.

Economic utility

Bauhinia leaves are large and flat and sewn together like banana leaves and made into dinner plates. Several articles like umbrellas, rain caps, rough table clothes, cups and small plates are made with dried leaves. The bark is made into rough ropes. The pods are roasted and the seeds are eaten by tribals. The outer bark contains about 17% tannin.

Place in bioaesthatic planning

The evergreen, elegant climbers can be planted near the large trees in public parks, colleges, near the school buildings to beautify the surroundings.

6.4 Copper pod or Rusty shield – Bearer (*Poltophorum pterocarpum* (DC) Backer)

Family: Caesalpiniaceae

Vernacular names

English: *Rusty shield bearer;* Tamil: *Ivalvagai, Perungondrai;* Telugu: *Kondachinta*

There is a gay with yellow mats of flowers in several cities during the month of April where rusty shield – bearer planted along the roads. This bright colour halo not only covers the lushy green ground the but also gives striking beauty to the roads. The fallen flowers covers the entire road like a golden carpet make the surroundings charged with beauty and tranquility. In some cities, *Rusty shield bearer* planted alternately with *Gulmohur* which also coms to flower at the same time. The brilliant yellow of rusty shield bearer and orange coloured *Gulmohur* make a best colourful combination and rise the spirits of nature lovers. In some areas the trees come into second burst of blooming at the start of monsoon and lasts before beginning of winter.

Geographical diversity

Rusty shield bearer is a native of Srilanka and it is distributed towards east upto Australia through Malayan peninsula. It was introduced in Calcutta botanic garden from Moluccas during the 19th Century. Interestingly the tree was spread all over the country within six decades. Being a good flowering tree, it was remarkably spread in Western India. The tree requires well drained soils. It is most commonly found in Bihar, West Bengal and Maharashtra.

Morphological features

Rusty shield bearer is a deciduous and grows to a height of about 15m. The branches are well spread and form thick canopy which can obstruct strong sunrays which results in cool. The scientific name *Peltophorum* means *"shield bearing"*, which refers to *shield shaped pods*. The species name *pterocarpum* means "with winged fruit". Sometimes the species is called *ferrugineum* means "rusty or reddish brown" coloured buds and fruits. On account of old copper coloured pod, the tree is also called '*copper pod*' tree. The copper coloured buds and pods are highly attractive with the thick green background leaves. The leaves are compound, feathery and bipinnate. About 10-15 pairs of leaf like pinnae are arranged on a central stalk. The leaflets are oblong, rounded or notched at the tip a closely set. They are bright green at early stage but later turned to deep green. The bark is smooth and grey. The tree blooms twice in the year, March to May and September to November. The handsome foliage and bright yellow flowers gives wonderful contrast and attracts the viewers. At first, buds appear on bare branchlets later formed into sprays. Petals are bright yellow with orange-yellow anthers. The stigma is like a miniature riveted shield or an umbrella. The flowers are slightly fragrant and arranged on large panicle. The blossoms are temporary and fall down with mild winds. A thick yellow floral carpet in formed on the ground. The pods remain for one year on parent tree.

Economic utility

The wood is used for making cabinets, tea boxes, planks and suitable for coach building. Leaves rich in protein (55%) and lopped for cattle fodder. Bark is rich in tannin and used for tanning leather, fish nets ad tarpaulines. Bark is used for dysentery and is constituent of tooth powders, and gargles in Ayurveda. Bark lotions used for sores, muscular pains and eye troubles.

Place in bioaesthatic planning

This highly ornamental tree can be planted in avenues, parks, gardens, railway platforms and vehicle parking places. When mixed with

Spathodea and *Gulmohar* the results are quiet conspicuous with peculiar colour combination.

6.5 Colvilles glory (*Colvillea racemosa* Bojer)

Family: Caesalpiniodeae

Vernacular names

Hindi: *Kilbili*; Telugu: *Yerrabatani, Turai*

Geographical distribution

The tree is a native of West Coast of Madagascar but some others claim that it's original home is East Africa. It was introduced into India as an ornamental tree during the foreign rule. Some of the beautiful avenues can be seen in the cities of Bangalore and Allahabad. The botanical name *Colvillea* is after the British botanist *Sir Charles Colville* and *racemosa* means *"flowers on raceme inflorescens"*.

Morphological features

Colville's glory is a medium sized tree with umbrella like canopy. It resembles like Peacock flower (Gold Mohur) in respect of stem and leaves but flowers are entirely different. The pinnate leaves composed from 20 to 30 pinnae. It usually flowers during rainy season (July to October). Flowers are orange to reddish in colour and arranged in large drooping racemes at the end of branches. They can be compared with grape bunches coloured with bright orange. Stamens are long and conspicuous. Scarlet fringe appears on flowers at the top of bunch. The flower buds are nut-like and very striking appearance. The floral clusters are highly attractive and the size of flower gradually decline at the end of receme. The fruit is a two walled rounded pod.

Economic utility

It is used as an ornamental plant in public gardens, hospital and factory compounds.

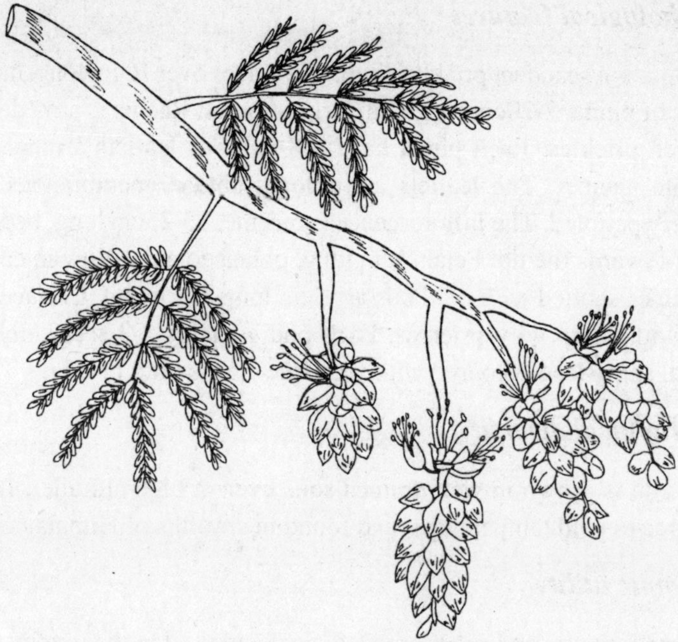

Fig. 6.1: Colville's glory (*Colvillea racemosa*), L – light green, F – orange to red, F – brown pod.

Place in bioaesthatic planning

The trees can be planted as avenues in city roads, public parks, Film studies, railway platforms where the moist climate prevails. It cannot withstand arid climate. The avenues are glorious sight when seen in full blooming. The tree should not be pruned before flowering as the inflorescens are present at the terminal branches.

6.6 The Fever-nut (*Cassia bonducella* L)

Family: Caesalpinia

Vernacular names

Hindi: *Kath Karanj*

Morphological features

Fever-nut is a scandent prickly shrub. The leaves over 10cm long, had 6-8 pairs of pinna, 7-20cm long with large stipules. Rachis is armed with recurved prickles. Each pinna have 6-8 pairs of leaflets arranged in opposite manner. The leaflets are oblong, obtuse, membranous and minutely petioled. The inflorescence is raceme, 15-25cm long, become denser towards the tip. Petals 1cm long, oblanceolate yellow in colour sometimes spotted with red Pods are 5cm long, dehiscent, Coriaceous, beset with sharp wiry prickles. Each pod contains 2-3 seed, globose shaped, shiny lead coloured with very fine cracks after drying.

Geographical diversity

The shrub is grown in well drained soils even at high altitudes. It has tolerance to cold temperatures and found in siwaliks of Himalayas.

Economic utility

The tender shoots and seeds are medicinal and used in the treatment of fevers. The farina of the cotyledons, whitish in colour, bitter in taste employed as good substitute for quinine. It is also used as hedge-plant to protect the crop from stray cattle.

6.7 Flame of the forest (*Butea monosperma* (Lan) Kuntze) (Syn. *B. frondosa koenig* ex Roxb)

Family: Fabaceae

Vernacular names

Bengali and Marati: *Palas;* Gujarati: *Khakra;* Hindi: *Dhak, Palas*; Kannada: *Mutthuga;* Malayalam: *Palas in samatha*; Sanskrit: *Palasha;* Tamil: *Parasa;* Telugu: *Moduga*

The palas is a native of India and frequently referred in sanskrit literature. In Ramayana, Valmiki described the Ravana's body with the flowering of palas tree when he wounded severely with Sri Rama's sharp arrows during the war in Lanka. Similarly Vedavyasa described wounded warriors with very striking blooms of Palas in Mahabharata war of *Kurushetra*.

If any one wish to see an impressive colour of Salmon-red flowers of palas, he has to travel in Western Railway from Delhi to Mumbai route *via* Kotah during the beginning of summers. One can appreciate the correctness of name after seeing the vivid mass of colour on the trees at the beginning of March. Similarly the traveller who is going to Delhi from Vijayawada, by train, can see the spectacular palas blooms in the dry deciduous forests of Madya Pradesh.

Geographical diversity

The Palas is found all over Indian subcontinent, Bangladesh and Myanmar. It mostly occurs in the forests of Madya Pradesh, Maharashtra and Andhra Pradesh. The tree also widely seen in sandy belts of Sourashtra and Gujarat from plains to mountains of 100m altitude.

Morphological features

Palas is a small to medium sized deciduous tree with crooked trunk and branches looking ugly except flowering period. It grows to a height of 6m and seldom rises beyond 7m. The leaves are trifoliate. The leaflets are leathery and the terminal one is stiff. The apex is rounded. The leaves hairless and smooth. The leaves fall off in cold season and flowers appear at the beginning of summer, that is, February and March. The flowers are large, with a size of human thumb. Theses flowers are densely crowded on leafless branches and gives an attractive fire like vermilon mass. If the entire forest is in full bloom, one can believe that forest was attacked with flames. The flower stalks are velvety to olive green in colour. The fruit is flat pod, 15-18cm long, narrowing at the apex in which single seed is enclosed.

Economic utility

As the palas tree looks wild and rugged, it seldom planted along the public roads and parks. The leaves are rich in nitrogen and used as greenleaf manure in wetland rice. In Konkan region of Maharashtra, the tree is lopped and branches are spread in the fields and burn them along with Cowdung cakes. The farmers believe that the ash will enrich the soil and improve the productivity of soil. Palas trees establish and succeed even in very poor soil. The tree used for the cultivation of lac insects in Bihar. Reddish juice may ooze from the cuttings of bark. The juice where exposed to air hardens into glassy ruby-red gum known as '*Bengal Kino*'. The kino is highly medicinal, being good astringent, helps in curing chronic diarrhoea. The seeds are employed against round worms and tape warms present in digestive canal.

The leaves are made into platters, cups, *bidi* wrappers, meals taking mats, and wrap up parcels. The flowers given an orange dye, and its decoction keep out white ants in the field. The wood is durable under water and used for well-curbs and piles. The bark of root yields a course fibre, used for making cordage and for caulking boasts on the river Ganges.

Place in bioaesthatic planning

The trees can be planted in rock gardens and borders of farm houses. The tree is beautiful with new leaves and flowers during April.

6.8 Gul Mohur or Gold Mohur (*Delonix regia* Rafin) Syn. *Poinaana regia* Bojer ex.Hook

Family: Caesalpiniaceae

Vernacular names

Adavi: *nippu chettu*; English: Flam-boyant flame tree, Royal peacock flower, Holy ghost tree; Goa: *Pentecost tree*; Hindi: *Gulmohur*; Tamil: *Mayarum*; Telugu: *Shima sankesula, Kodipandela chettu*

The original home of the tree is Madagascar and introduced into Mauritius and India at the start of the 19th century. Britishers planted this tree in Royal Botanic garden of Calcutta around 1840. Later it became one of the Commonest tree in several urban parts of India. The name *Delonix* means *"with a claw or nail"* as the shape of the petal resembles that of tiger claw. In hindi speaking belt, it is called as *Gulmor*. The word *Gul* means *'rose'* or *'flower'* and *Mor* is hindi name for *'peacock'* which is known for its beautifulness. Some times it is called as *peacock rose*. The christians of West Coast are referred the tree as pentecost or *Holyghost* tree as its peak flowering occurs around the month of feast of pentecost.

Geographical diversity

Gul Mohur is mostly grown in tropical and subtropical regions. The tree is well adapted to a variety of climate and soil conditions. It is grown in high rainfall areas and in semi arid regions of the country as well. The tree can not tolerate severe winter and frosting. The tree can do well on a rocky soil.

Morphological features

The Gul Mohur is a medium sized tree, attains to a height of 8-10m. The growth of tree mainly depends upon nature of soil and climate. The tree can easily be blown down during typhoons and storms usually occurs in monsoon season because of its shallow root system. The tree is a fast grower and produce umbrella like canopy. The leaves are green and of very pleasant hue. The leaves are compound and has 50cm length. The leaves consists of 10-20 pairs of pinna on a midrib. Each pinna bears 20-30 pairs of small green leaflets. The tree shed its leaves before flowering stage, about the end of February. When leaves fall-off completely, flowers appear in large number. When the whole tree crowded with bright red flowers and appears like fire tree. Flowers are arranged on large racemes at the end of branches. The flower consists of 4 bright coloured scarlet petals. The fifth standard petal, larger in size and variegated red and yellow in colour. The petals are spoon shaped with crinkled margins, appears like a claw of wild animal. The fruit is a long, flat pod. The fruit turns deep brown and grows to a length of 0.5m. The seeds are oblong and mottled.

Economic utility

As it most showy and can be planted on roadsides in villages and towns. Seeds contain gum which is used in textile industry and food industries. The wood is white and light. The wood takes a fine polish and used for making combs, matches and churns.

Place in bioaesthatic planning

The Gul Mohur is very suitable for avenues where both flowers and thick shade are desired.

6.9 Indian Coral tree (*Erythrina variegata,* orientalis(L) Meril)

Family: Fabaceae

Vernacular names

Bengali: *Palita mandar;* Gujarati: *Bangaro, Panaraweo;* Hindi: *Dadap, mandara;* Kannada: *Harivana, varjipe;* Malayalam: *Kalayana murikku, Mandaram*; Marati: *Pangara*; Sanskrit: *Mandar, Parijata;* Tamil: *Kaliyana murukku*; Telugu: *Badisa, modugau, baridamu, badita*

Folk lore

According to Hindu mythology, the tree belongs to heaven and grown in Devendra's garden, who is the king of *suras* and angels. The bright scarlet coloured large flowers attracted *Satyabhama*, the dearest wife of Sri Krishna. She requested her husband to bring the tree to the earth from heaven. Sri Krishna attacked Devendra defeated him and brought the tree to the earth and planted in Satyabhama's garden. Hindus believe that the leaf having three leaflets represent *Trimurti*, i.e. the left Shiva, middle *maha Vishnu* and right side lies *Brahma*. It is like Shamrock of

Ireland and Christians believe that three leaflets represents the disciplies of St. Patrick.

Geographical diversity

The name, *Erythrina*, has been derived from Greek word *Erythros* meaning coral red, refering to the red coloured blooms and *indica* refers to Indian origin. Indian coral tree is one of the highly colourful indigenous trees of tropical India especially along the sea coasts. In north India the tree was mostly found in Bihar and West Bengal. It also spread into Andaman and Nicobar islands, Myanmar, Indonesia and Polynesia and in the sub-Himalayas.

Morphological features

Indian coral is a small to fair-sized tree but quick growing. Under fertile and well textured soils, the tree attains to a height of more than 15m. The branches have a closed frame work and armed with very sharp prickles. The bark on the trunk is smooth and variegated trifoliate leaves gives much elegance to the tree. The leaves are egg shaped. The branches are fairly straight and gives rise to floral spikes at the very end. The large scarlet red, pea shaped flowers borne in racemes, arranged in cluster, appears like multifingered hand. Before flowering, the tree completely shed its leaves towards the end of February when atmosphere warming up. Flowering continues upto the end of April. The colour of flower is bright red. When the flowers open out, the calyx splits the whole way down to the base. The flowers persists on the tree for 4 to 8 weeks.

The pods are 15-30cm long, green at first but turned to deep brown and contains nearly 12 seeds. The pod constructed between the seeds. The fruits may remain on the tree as long as the leaves dries off. The tree can be propagated either through seed or through branch cuttings.

Economic utility

The tree is known for its beauty and grown as ornament in houses. It is cultivated as a support for various climbers like grape vines, betel leaf, gourds, pumpkins, creeping roses and Jasmines. Coral trees are grown in the tea plantations to enrich the soil with nitrogen. It also acts as shade tree in betel vine. The trees are used as hedge plants as wind break in horticultural gardens. The young leaves are used for making curries. The wood is used for making carved articles.

Place in bioaesthatic planning

The trees when planted in multiple rows along the roads, creates a spectacular beauty to surroundings during the flowering period. Good avenues can be seen in Chandigarh and one can breathless when the tree in full bloom.

6.10 Java pink cassia (*Cassia Javanica* L)

Family: Caesalpiniaceae

Vernacular names

Hindi: *Java-Ki, Rani*; Telugu-*Java, Tangedu*

Geographical diversity

Java cassia is most beautiful tree among cassias and indigenous of Malaysia and Java. It was well spread throughout India. As the tree is somewhat brittle, it is liable to break against strong winds and typhoons which are common in coastal areas. In sheltered situation, it grows well and spread its crown extensively. Good growth can be seen in well aerated and deep soils.

Morphological features

Java cassia is a medium sized handsome tree with straight trunk and spreading crown. The branches spread horizontally with droopy leaves. The leaves are feathery and bright green in colour. The leaflets are bluntly rounded at the apex. In hot summers, a new flush of leaves borne on the branches. The green flush accompanied by deep-pink buds in clusters at the end of branchlets. The flowers are grouped in whorls and highly attractive. The loosely borne floral clusters can be seen for a short period during May and June. It is a goodly sight to see the scene of new bright green leaves intermingling with pinkish floral clusters.

Economic utility

Wood is used for tool-handles, ploughs, wheels and house posts. The bark can be used as tanning material in leather processing.

Place in bioaesthatic planning

The handsome tree can be grown as avenues in metropolitan cities. These avenues are more common in Mysore, Chandigarh and Bangalore.

6.11 Kadam (*Anthocephalus Chinensis* Walp)

Family: Rubiaceae

Vernacular names

Kannada: *Kadawala*; Malayalam: *Attutak*; Sanskrit, Hindi, Marati, Gujarati: *Kadamba*; Tamil: *Vellai Cadamba*; Telugu: *Kadambum*; Trade: *Kadam*

Folklore

Kadamb is indigenous to Indian sub-continent and references are made in ancient sanskrit literature. The tree was frequently referred in *Sri Maha Bhagavata* when described thc places around Yamuna river. It is said Sri Krishna played several melodious songs under the Kadamb tree with his flute. SriKrishna used to meet his ardent lover, Radha under the shade of Kadamb tree during his boyhood. When Sri Krishna left to Magadha from Repalle, to kill her uncle King Kamsa, Radha used to stand under the kadamb tree awaiting to her lover's arrival. It is described that Radha encircle the tree with her hands with unbearable pain of Sri Krishna's separation. Interestingly thick stands of Kadamb trees are found in Barsana, the birth place of Radha even now. Hindus considered the tree is highly sacred as it provided cool shade to

Bhagawan Sri Krishna. Some devotees believe that the tree is mute form of goddess *Sri Kali* and worship nearly ten days during *Dushara* festival.

Geographical diversity

Kadamb is a native of deciduous forests of India. In Central India, the tree is considered as precious wood yielding tree after teak. The tree grows in moist soils provided with good drainage. At seeding stage, it can not tolerate strong sunlight and heavy rains. The seedlings are highly sensitive to water stagnation. Besides in Indian sub-continent, Kadamb was also found in warmer areas of China and Malaysia. The largest trees of height of more than 55m with a girth of 6m can be seen in Drang forests of Maharashtra.

Morphological features

Kadamb is a huge deciduous tree grows over 30m. leaves are broadly oval or circular in shape and sharp of the apex, heart shaped at the base. Young leaves are green tinged with pink colour with strong nerves. Leaves are borne in pairs with stipules. Flowers are yellow and very small and insignificant individually. When little flowers open out, these are like flower balls. Fruits are small, forms on solid ball, turns to black during ripening. The tree shed its leaves in February and remain bare upto end of March. Flower balls are formed at the beginning of monsoon and beautify the surroundings.

Economic utility

Flower receptecles edible. Bark is medicinal and used as tonic and febrifuge. Flowers yield an essential oil which is used in confectionary and sweetmeats. Wood is used for the manufacture of tea boxes, packing cases, for boarding and also employed for house construction work as rafters and beams. The wood is also suitable for making canoes, dugouts, carving and turnery. Pulp can also be used for making paper. The bark is antiseptic and tree juice is used for killing worms in sores. Flowers yield an oil which is used in perfumary. Leaf decoction is used for gargling the throat.

Fig. 18:
Covilles glory (*Colvillea racemosa*)

Fig. 19:
Gold mohar (*Delonix regia*)

Place in bioaesthatic planning

Kadamb can be planted as avenues as the tree is showy when in leaf. The tree is a delight to of the surroundings. The yellow flowers are prettiest when seen against the blue of sky. The fragrance of flowers resemble the fresh fine wine and raise the spirit of nature during cloudy monsoon.

6.12 Lignum Vitae (*Guaiacum officinale* L)

Family: Zygophyliaceae

Vernacular names

English: *Gum Guaiacum, Tree of life, Brazil wood*

Geographical diversity

Lignum vitae means the '*tree of life*' or life giving tree most possibly due to high medicinal value. Although the tree is a native of West Indies, it was well distributed in Southern states of India. Excellent specimens can be seen in chennai, Mumbai, Bangalore and Bhagyanagar. It thrives well under moist soils coupled with warm temperatures.

Morphological features

Lignum vitae is a small evergreen tree with crooked stem and with deeply furrowed bark. The leaves are compound with dark-green leaflets. The leaflets are stalkless arranged in 2-3 pairs, on thin midrib. The leaflets are very smooth and gracious. The flowers are bright blue and appears in clusters. The tree blooms in November and March. The tree in full bloom is a goodly sight because of their colour contrast with dark green leaves. The floral colour fades from deep to light blue after a few days of flowering. The green turned into yellowish on ripening. It can be propagated through seed.

Economic utility

The wood is heavy, hard and used for making bowls, rulers, pulleys and ships blocks. Because of its sinking nature, the wood can not be used for boat making. The bark yields some kind of resin which is used in Ayurvedic medicines. Gum Guiacum is used as fat stablized in paints and varnishes. It is a mild laxative. It is also employed for curing tonsilitis, and pharyngitis.

Wood is used in the manufacture of segment bearings in steamship propeller shaft assemblies. It is self lubricating property, resistance to salt water and great compression strength. Also used for making stencil, and Chisel blocks, pestcles, bowls, pulley sheaves, brush beaks, skittle balls that require great strength and durability.

Place in bioaesthatic planning

The tree can be planted in all public parks, private gardens, frontyard of star hotels and guest houses because of its variegated pattern of colours with leaves and flowers.

6.13 Mexican Silk cotton (*Chorisia speciosa* St. Hill)

Family: Malvaceae

Vernacular names

Telugu: *Mexico Boorugu Chettu*

Geographical diversity

Mexican silkcotton is a warm loving tree and grown in tropical regions of India. The tree was found abundant in Southern states like Kerala and Tamilnadu. The tree needs fertile nutrient rich soils and abundant moisture. Luxurious growth can be seen in moist high rainfall regions.

Fig. 20:
Mixican silk cotton
(Chorisia rpeciosa)

- green stem
F - green
Fl - light yellow

Fig. 21:
Rhododendron
(*Rhododendron arboreum*)

L - Dark green
F - Red

Mosphological features

The mexico silk cotton is a fast growing tall deciduous tree. It begins to flower after five years. The trunk is bottle shaped and green in colour. It propagates through seed and germinate during monsoon season. It flowers before the advent of winter (October). Before flowering, the tree completely become leafless and bare. The flowers are light yellow and highly attractive against the background of sky blue. It bears numerous flowers and gives glamour to the surroundings.

Economic utility

The wood can be used for making match boxes and sticks. It is also usedful in packing box industry. The cotton can be used for stuffing the matteresses. The gum oozing from stem wounds used for binding the paper.

Place in bioaesthatic planning

The trees can be planted on both sides of roads in towns, villages and in parks. They can be used in formal planting schemes in metropolis.

6.14 Mauve Tabebuia (*Tabebuia speiosa*) Syn. *Tabebuia serratifolia* (Vahl) Nichols

Family: Bignoniaceae

Vernacular names

Telugu: *Garata puvvu chettu;* Trumpet flower tree
Brazil: *Tabebuia*

Geographical diversity

Tabebuia is a native of Mexico and mostly grown in tropics and subtropics. It prefers moist soils which are rich in fertility.

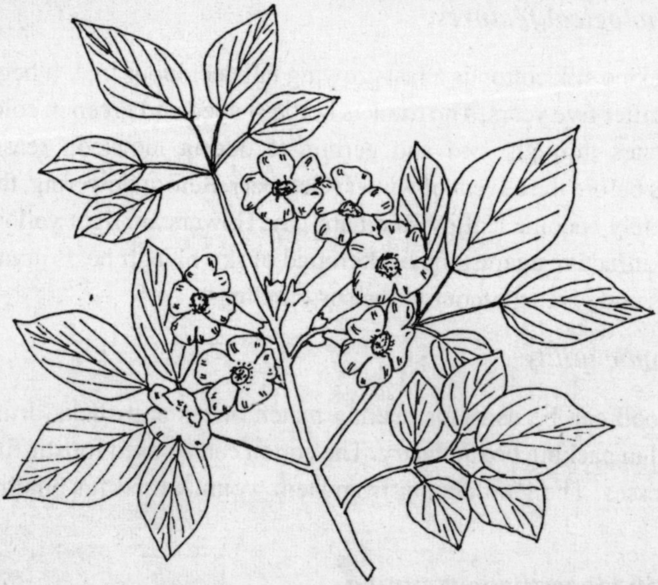

Fig 6.2: Mauve tabebula (*Tabebuia speciosa*), L – green, Fl – pale mauve, Fr – dark brown.

Morphological features

Tabebuia is a small erect tree with deciduous nature. The leaves are compound and palmately five leaved. The flowers are large, funnel shaped arranged on a common peduncle. Flowers arranged in groups in a beautiful manner. The flowers are pale mauve in colour and highly handsome. The tree is fully bare when blooms in june. The fruit is a capsule with undeciduous style at the tip. The seeds are winged and, flat and desseminate through wind.

Economic utility

Tabebuia wood is suitable for making house- framing, walking sticks, tool handles, archery bows, fishing rods and cross ties. Some are made into veneers for decorative work.

Place in bioaesthatic planning

The tree can be planted on the side of roads, foot paths, industrial road paths, public parks, hospitals for its decorative flowers.

6.15 Mussaenda (*Mussaenda frondosa* L)

Family: Rubiaceae

Vernacular names

Bengali: *Nagbali*; Hindi: *Pedina,Nagavali*; Kannada: *Billoothi, Pathri;* Lepcha: *Tumberh*; Malayalam: *Paratholi*; Marati: *Bhutkesa, Bhutkas*; Nepali: *Asari*; Sanskrit: *Srivati*; Tamil: *Vellaiyilai*

Geographical diversity

Mussaenda is a native of Northern India and the trees are found wild at the foot hills of Himalayan region. The trees are also found all over the Indian peninsula particularly in Western Ghats. The tree prefers fertile soils with high moisture holding capacity. Whilst travelling through the Nilgiris and Darjeeling, one can see the groves of mussaenda shrubs with the coloured leaves. The young leaves are thick green, old leaves are yellowish and large snow white or creamy white leaf near the flowers. The white leaf is actually not a leaf but modified form of calyx. Enlargement of sepals as white calyx is thc peculiarity of the tree and gives special look among the flowring trees. The tree can be grown as special look among the flowering trees. The tree can be grown even in poor soils when moisture is sufficient in the soil.

Morphological features

Mussenda is small to medium sized tree, attains to a height of 6-8m. The leaves are elliptic and opposite. Hairs present on the upper surface and nerves are seen beneath the leaf. Leaf base and apex are tapering. The leaves are deep green. Inflorescence is like large loose bunches. The flowers are orange colour inside and greyish outside with fairyl large and open cymes. One of the sepals of calyx, outside of the cyme, begins to enlarge into a leaf like structure and turned into normal leaf size. This calyx leaf is creamy white and later turned to dirty white and persists for a long time.

The tree flowers at the end of monsoon but the white calycine leaf can be seen throughout the year. The fruit is green pear shaped and minute seeds are immersed in reddish spongy tissue.

Economic utility

People believe that mussaenda is a favourable tree of '*Sri Mahalakshmi*', the goddess of fortune. The flowers are used for worshipping the Lord Vishnu and Srikrishna. The flowers made into garlands and tied over the doorways on festive occasions.

The leaves and foliaceous calycine lobes of flowers are eaten as vegetables. Leaves are used as green manure in field crops. The whole plant is important as it is medicinally used very often in rural areas. The juice of leaves and fruits are used as an eye wash and believe that it removes opacities. Decoction of dried shoots was given to children for curing the coughs. Leaf decoction is good medicine for killing the intestinal worms. Roots are demulcent and used for white leprosy and eye troubles. Leaf paste is applied in case of boils and skin ulcers. Wood is used for making small articles like cooking spoons, ladles and tool handles.

Place in bioaesthatic planning

The trees can be planted in hotels, dak bungalows, village schools and near the ancient buildings and ruins under Archaeological Department for shading and to add colour to the surroundings.

6.16 Mountain Ebony (*Bauhinia purpurea*, Linn)

Family: Caesalpinioideae

Vernacular names

Bengali: *Devakanchan, Koiral, rakta kanchan*; Dun: *Khairwal*; English: *purple bauhinia*; Garhwal: *Guiral*; Hindi: *gulabi kachnar*; Kannada: *basavanapadu*; Marati: *atamtti*; Oriya: *boroda*; Saharanpur: *Hairwal papri*; Tamil: *Kalavilaichi, mandari*; Telugu: *Kanchanam, bondanta*

Geographical diversity

Purple bauhinia occurs in wild all over Indian sub-continent. It is very commonly found along the foot-hills of Himalayas from Indus to Brahmaputra region. The tree can also be found extensively in all hill ranges of Central and Peninsular India. It is grown on a large scale as an avenue in all hill towns. The tree named after two botanists, *John*, and *caspar bauhinea* and *purpurea* means "purple red" coloured flowers. It flowerishes in well-drained soils and advessely affected by cold temperatures.

Morphological features

Purple bauhinia is a moderate sized evergreen tree, densely branched with rounded crown. The bark is dark-brown or ashy-grey and smooth. The great sanskrit poet, Kalidasa referred this tree as *Kovidara* in the drama of *Meghsandes*. Leaves are simple, bilobed, cordate, 9-11 nerved, less frequently truncate, coriaceous, cleft about half-way down, giving a hoof like appearance. This character makes the *bauhinia* easily identified. The buds are fine-angled, dark green and downy. Flowers deep-rose, the calyx splits into two reflexed segments, one having two wings and other three on opening. Flowers are purple to lilac, large, showy and five-petaled. The base of one petal is white. The tree flowers profusely at the beginning of winter (November). The pods are 18-25cm long, Sickle like. The green pods, looking like french beans among the leaves. The pods are pubscent, greenish-purple when young, late in dehiscing and contains 12-15 seeds. The tree flowers from September to November and fruiting occurs from January to March.

Economic utility

The bark is used for tanning the animal hides. The flower buds are cooked and eaten as vegetable. Sometimes the buds are pickled. The leaves are lopped for cattle particularly for camels and horses. The heartwood is moderately hard, reddish-brown, fairly close-grained and used for making farm implements and house construction.

The bark of underground root is poisonous and used as a rat bait.

Place in bioaesthatic planning

The tree can be planted as road-side avenues in towns and cities on account of the fact that it is one of the few trees flowers in cold season. It is admirably suited for small gardens, Front yards of houses, public parks and office buildings. Rural people usually cultivated the tree for its most beautiful flowers.

6.17 Queen Crape Myrtle (*Lagerstroemia speciosa* Pers)

Family: Lythra ceae

Vernacular names

Assamesse: *Ajhar, thing-donthlado*, ; Bengali: *Ajar,Jarool*; English: *Pride of India*, Queen's flower; Hindi: *Jarul, arjuna*; Kannada: *hole dasavala Challa, poledachalla*; Oriya: *Patoli, ary*; Punjabi: *Jarul*; Tamil: *Kadati, pumaruthu*; Telugu: *Varagogu*; Trade: *Jarul, Pyinma*

Geographical diversity

Queen's flower tree grows best in fertile deep alluvial loams. It required moderate sunlight but growth depression takes place in heavy shady areas. The tree prefers moist soils and found in small forms in dry areas. It grows to a large size in moist deciduous forests of peninsular India but smaller size in dry areas of Northern India. The tree is widely distributed all over India, but the true home seems to be forests of eastern India, Myanmar, and Sri Lanka. Luxurious growth was also seen in Kerala and Assam forests. The tree is very common on the banks of rivers and near the lakes and swamps. The botanical name *Lagerstroemia* is named after Swedish botanist, *M.V. Lagerstrom* and *flos-regina* means '*Flower of Queen*'.

Morphological features

Queen myrtle is medium sized tree but small in dry areas and attains to a height of 6 to 10m high. It has a short bole, rounded crown with light-gray bark. The undersurface of leaves are veined densely and pale in colour. The tree rarely seen bare as the leaf fall is slow. At the end of winter (February – March), the leaves arises in hot months (April-May). The inflorescence is terminal but flowers are mauve purple or pinkish at initial stages but turned to white before fall on the ground. These flowers are highly showy and gives festival to eyes during hot months (April to August). The fruit is woody capsule, having 5-6 valves. The fruits ripen during winter and shed light-brown seeds. Seeds have stiff wings which helps in dispersal by wind. At the early stage, pruning of side branches is necessary for training a well-shaped tree.

Economic utility

The timber of tree is valued next to teak and used for planking, ship building, bridges, water tanks, piles, boats, dugouts, oars, railway carriages, mortars, careel pounders, mine props and motor lorry bodies. The wood is also employed for making leather cutting blocks, boot lasts, matchsplints, railway-sleepers, telegraphic poles, wheels, carts, boxes and other furniture.

Leaves are purgative, diuretic and deobstruent. Roots are astringent. Seeds are considered to be narcotic. Bark is also used as purgative in native medicine. Tanin extract is prepared from fruits and leaves. The fruit is used for curing ulcers in the mouth in Andaman islands.

Decoction of leaves and fruits are employed in diabetes treatment. A fibre can be extracted from bark.

Place in bioaesthatic planning

Queen flower tree is well known avenue tree. The flowers are highly beautiful. The trees can be planted on the side of roads in cities and towns. The avenues can be seen in lucknow and Chandigarh.

6.18 Rain tree or Sleepy ganer tree (*Enterolobium saman prain*) (Syn. *Pitheocolobium saman* Merril)

Family: Mimosoideae

Vernacular names

Hindi & Bengali: *Belaiti siris;* Malayalam: *plavu;* Tamil: *amaivagai;* Telugu: *Nidra ganneru chettu*

Geographical diversity

Rain tree is a native of Central America Voyagers. The tree is spread rapidly throughout tropical parts of Indian subcontinent. The trees are abundant in West Bengal. Andhra Pradesh, Orissa, Tamilnadu and Kerala. As the tree is very common on railway platforms for shadi purpose, people call it as railway tree. The tree requires warm weather and high rainfall. In Indo-Gangetic region, the trees liable to be killed by frost during Chilling winters. The leaves become droopy during heavy rains hence it is termed as rain sensitive tree.

Morphological features

Rain tree is a large, quick growing. handsome tree with a broad spreading crown. The tree is partially deciduous but maintains canopy evergreen. The foliage is feathery with pinnate dark green coloured leaves. The bark is coarse and fissures are seen. The trunk girth sometimes reach upto 3-4 meters but the wood is not hard. Sometimes the trunk wood is damaged by insects and big tunnels are formed. Honey bees make the hives in the tunnels. Parrots and wood peckers also make the small holes on the trunk and their eggs in them. Some moisture is always present under the tree which attributed to the discharge of Cicadas feeding on the green leaves of tree. The tree sensitive to solar radiation and light. On bright sunny day, the leaves spread horizontally and not allow the sunlight through the dense canopy. However, during night, or overcast sky or rainy weather, the leaves become droopy and folded sideways.

The leaves are pinnate, long, heavy and has no stalks and curved towards the end. The flowers are pale-pink and appears in large clusters at the end of branches. The flowers look like silken tufts. The tree flowers at the end of winter and continue upto end of South-West monsoon (March to September). The fruit is fleshy pod, indehiscent and liked by squirrels and rodents. The tree can be propagated through vegetative cuttings or seeds. Growth is very rapid.

Economic utility

The sweet pulp of pod is relished by buffaloes, Goats and cattle. People say that the buffaloes and cows yield more milk and butter when the pods fed to them. Leaves and pods used as fodder. The tender leaves are relished by cattle. Tree yields a gum which swells in water into a tough cartilage like mass. Like carob beans (*Ceratonia siligua* L) they contain a sugary pulp which relished by horses. As the tree grows to a gigantic size, honey bees prefer to makc beehives on the top of branches.

Place in bioaesthetic planning

The tree is not suitable in the compounds of small houses and Dak Banglas because of its spreading habit and occupy more areas. It can be recommended to plant along the roads, canal banks and high-ways because of its excellency in providing cool shade.

6.19 Rhododendron (*Rhododendron arboreum*, Smith)

Family: Ericaceae

Vernacular names

English: *Tree-Rhododendron*; Hindi: *Burans, Buras*

Geographical diversity

Tree rhododendron was mostly found in Himalayas particularly in Simla hills, Tehri-garhwal ranges at an altitude of 1500m-3500m. It was frequently associated with *Quercus incana* and *Pieris ovalifolia*. The tree prefer acidic soils.

Morphological features

Rhododendron is a small evergreen tree and attains to a height of 10-15m. The bark is reddish-brown and peeling-off in small flakes. The leaves are crowded at the end of branches, 10-16cm long, lanceolate or elliptic-oblong, acute at both ends. The leaves are glabrous above, rusty tomentose or silvery beneath. The midrib and nerves are prominent beneath. The buds are viscous. Flowers are large, in dense clusters, very showy with deep-crimson colour, sometimes in white and pink. The flowers are arranged in Corymbose fascilles at the end of branches. The pedicels are short with minute calyx and companulate corolla. It flowers in the Himalayas from March to May when the air warms up. Fruiting occurs in Autumn and Winter.

Economic utility

The large flowers are acid in taste and are used for making Chutney, pickles and preserves. Sometimes they are eaten directly by the hill people. They also offered to Gods in temples. Hill people believe that the flowers are good remedy for dysentery and are applied as a poultice in headache. The viscid tender leaves are eaten cooked as a vegetable in mountane areas.

The wood is brownish or reddish-white, soft, close and even grained, seasons badly and is apt to warp and shrink. It occasionally used for building houses, for making dishes, gun-stocks, tool-handles and other agricultural implements. The wood is chiefly used for fuel and charcoal making.

Place in bioaesthatic planning

The trees can be grown as avenues in small roads, public parks and slopy lands. In warm season, the entire area turned to scarlet gleam with blooms of rhododendron.

6.20 Silk-cotton or Cotton tree (*Bombax Ceiba* L)

Family: Bombacaceae

Vernacular names

Garo: *boichu*; Gujarati: *Sawar, simalo,shemolo*; Hindi: *Semul, shembul*; Malayalam: *mullilavau*; Marati: *Saur, simlo, samar,kantesaver, shevari*; Mundari: Edelsong.; Oriya: *Bouro*; Rakat: *senbal,Kaanti senbal, pagun*; Rakta: *Puspha, Kantakadruma,*; Sanskrit: *Salmali, Shalmali*; Tamil: *mullilavu, illavam*; Telugu: *Konda-buruga, mundala burugu*

Folklore

According to Hindu mythology *pitamaha Brahmadev* taken rest under the shade of Simul tree after the creation of living world. Hence, the tree has been seen with reverence. The tree has been found frequently in sanskrit literature on the name of *Salmalia* which conjures up the world of traditional world. The lofty tree is also much celebrated in the classical poetry of hindi and English. The poets felt excited for its spread of large opulent purplish-red flowers when in full bloom. Several authors felt that there are few trees which can take an account on fingers which are on par with glorious display put up by *simul*. The flowers of *simul* is usually crimson red but occassionally some tree bear yellow and creamy White flowers. Rural folk believe that the yellow flowered tree is having the miraculous powers of divine. The devotees who practice or trying to get benefits from evil spirits, do not touch the tree. The tree attracts several birds like Mynas, rosypastors, coloured sparrows which chatter and singing continuously during spring season. No human can excite and stunned by seeing such beautiful scene. Even monkeys climb the branches in groups to draw the nectare from the thick petals of flower. Squirrels are very fond of these sweet petals and move on the branches speedily for eating the flowers. If one can have the *simul* tree near the bed room, he will be covered with cotton masses when he awaken from sleep in the dawn. The broken matured fruits release the cotton fibres by even mild breezes and spread the surroundings.

Geographical diversity

Simul is a native of Indian subcontinent and occurs all over the country from Garhwal to Deccan peninsule. The tree also frequently found in Western sector of sub-Himalayas, Bundelkhand, Bihar, Orissa, Madya Pradesh and West Bengal. The trees are grown luxuriously from plains to 1200m. altitude. The tree prefers hot weather, dry, well drained stony and gravel regions. The tree also frequently found in Myanmar, Bangladesh and Islands of Eastern Archipelago.

Morphological features

The simul is a tall deciduous tree with numerous spreading branches. The tree often reaches more than 25m. height and Characterized by stout trunk. The bark is smooth and ash, coloured. At young stage, the bark is covered with sharp conical prickles which helps in keeping away of stray animals. The leaves are large, bright green, and stiff in texture. Each compound leaf consists of 5-7 leaflets. These leaflets are lanceshaped, acute or tapering at the ends. The leaves remain green for the greater part of the year and make them one of the best trees in the world. The leaves fall-off before flowering at the end of January an flowering starts February onwards. The flowers are bright crimson coloured. The flowers borne on a large number and the tree appears covered with flower mass. The calyx is thick cup shaped. Fruits are large

Fig. 22:
Red silk cotton tree
(Bombax ceiba)

Fig. 23:
Scarlet cordia *(Cordia sebestena)*
Bark - Brown
L - Green dark
Fl - Brilliant orange red
Fr - White berry

Fig. 24:
White Frangipani (*Plumeria alba*)
L - Dark green
Fl - All white

capsules, reddish at initial stages and contain kidney shaped seeds that are embeded in silky threads.

Economic utility

Simul is highly valuable as every part of tree is useful. Its transparent gum is a substitute for *tragecantu gum* which is used by shoe makers. Tender leaves are eaten and lopped for fodder. Flower buds and fleshy calyses eaten as vegetables. Bark is demulcent, tonic and styptic. Bark is demulcent, tonic and styptic. Fruits are expectorant, diuretic, stimulant, used in calculus affections and ulceration of bladders and kidneys. The *Mocharus gum* oozing from tree is demulcent and styptic. Seeds yield an edible fatty oil which is used for soap making and as illuminant seeds are edible and are eaten when roast. The floss covering the seeds is like Java kapok for stuffing mattresses, pillows, cushions and life belts. The decoction of wood is nutritious and used as medicine.

The wood is soft and used for making sticks, match boxes, shingles, canoes, toys, scabbards, cooperage, bush handles, well curbs, tea-chest, plywood, pencils and pen and sound-proof covers. Fibre is spun into yarn used for the manufacture of plushes.

Place in bioaesthatic planning

The beauty and brightness of flowers and shadiness of leaves make it as best avenue tree. It can be planted along the streets of cities and along the national highways.

6.21 Scarlet Cordia (*Cordia dichotoma,* Forst F)

Family: Boraginaceae

Vernacular names

Bentali: *Bahubara*; English: *Aloes wood*; Gujarati: *Bargund;* Hindi: *LasoraLal lasora*; Kannada: *Chikkachalle*; Kannada: *Chikkachell*; Malayalam: *Viri, Cheruviri;* Marati: *Shelvant*, Sanskrit: *Bahuvaraka*; Tamil: *Naruvili*; Telugu: *Chinna nakkeru Virigi*

Geographical diversity

The original home of scarlet cordia seems to be cuba. The tree was well distributed throughout tropics. It also found both in Indo-Gangetic plains and peninsular India. In Himalayas, It occurs and spread upto 2000m. elevation. The tree requires rich and well drained soils.

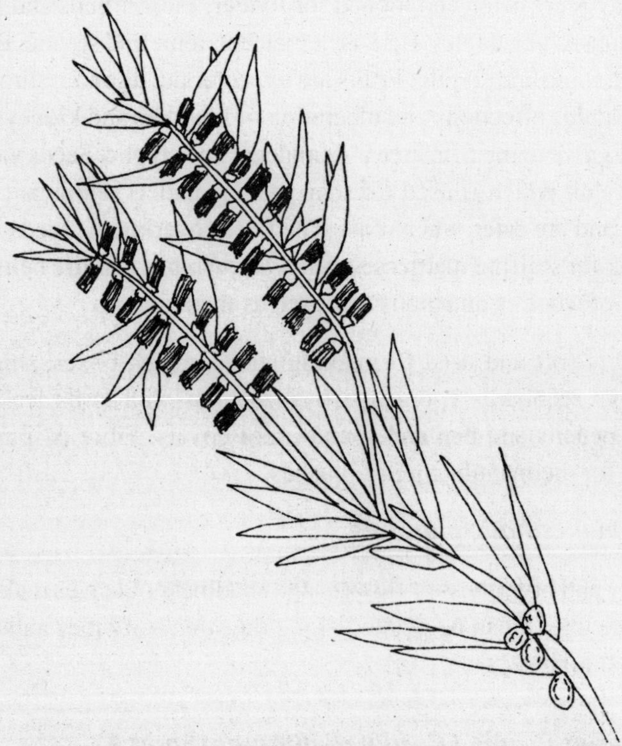

Fig. 6.3: Scarlet bottle brush tree (*Callistemon lanceolatus*), L – green, Fl – scarlet red, Fr – dark brown.

Morphological features

Scarlet cordia is a small deciduous tree and grows upto 20m in native forests but in gardens it become a shrub of 12m high. The trunk is short and crooked, covered with brown bark having small ridges. The leaves are egg shaped, large with blunt tip. The veins are prominent and deeply drawn. The leaves are dark-green wrinkled and coarse textured like rough sand paper.

The inflorescence is large open clusters borne at the tip of branches. The flowers are brilliant orange red sometimes deep scarlet, presents a rich hue during flowering in winter. In general, the tree flowers all the seasons but rich blossoms can be seen from January to March. The fruit is white berry consists of seeds embedded in sticky pulp.

Economic utility

The fruit of scarlet cordia is known as Sebestan which is edible and eaten by rural people. The fruit extract is used as astringent, diuretic, anthelmintic, demulcent and expectorent in country medicine. It is widely used in the diseases of urinary and chest pains. The wood is water resistant and durable in contact water. That is why, it was employed to manufacture boats, well-curbs and agricultural implements.

Place in bioaesthatic planning

Cordia is highly valued as a decorative tree and can be planted as avenues in University roads, medium roads of towns and villages. It also suitable for beautification of public parks, sports stadiums and star hotels.

6.22 Tulip tree (*Spathodea campanulata,* Beauv)

Family: Bignoniaceae

Vernacular names

English: *Scarlet fountain, syringe tree, Uganda Flame tree, Bell; Flambeau tree, African tulip, Squirt tree.*; Hindi: *Rugtoora*; Kannada: *neerukaye, Lujjekaye*; Tamil: *Patadi*; Telugu: *patade, patadiya*

Geographical diversity

Tulip tree is native of East Africa and distributed in all tropical countries. In India, it is widely spread from Punjab to Peninsular India. The tree prefer highly fertile soil and demands good drainage. Tulip thrives well from plains to 150m altitude. It requires moderate rainfall (50 to 100cms) for its luxurious growth. The tree is liable to break and fell due to strong winds.

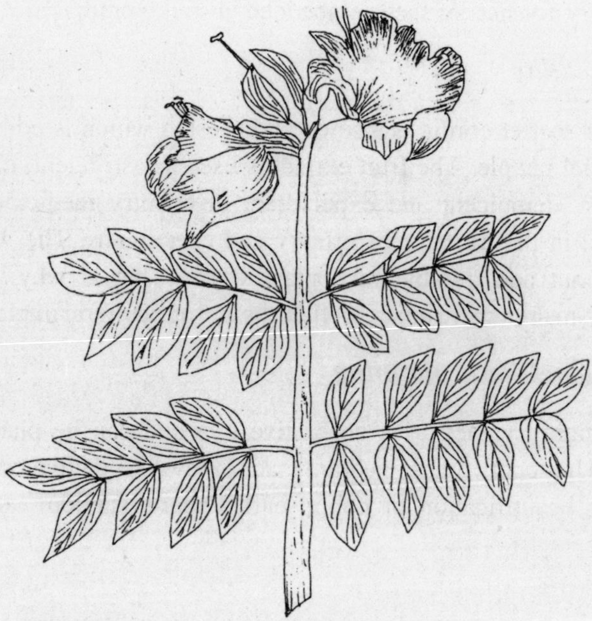

Fig. 6.4: Tulip (*Spathodea companulate*).

Morphological features

The tulip is a medium sized, slender, closely branched with dark green foliage. The bark has deep brown colour. The leaves are massed at the tip of branches. The leaves are coarse and pinna shaped. The leaves are egg shaped and prominently veined. Shoots and under side of young leaves are velvety. This deciduous tree shed its leaves at the and of winter in dry regions. The large flowers are bright orange-crimson coloured globe like shape. The inflorescence is cluster shaped. Inflorescence arises at the tip of top branches.

The botanical name *Spathodea* is derived from Greek maeans *"spathe"* refers to the *'spathe like calyx'* and *campanulata* is a stout leathery calyx filled with liquid. The liquid is absorbed by flower before it opens petals. The tree flowers pointed at the tips. The pods are like fingers pointing upwards above the green leaves. Seeds are papery and white.

Economic utility

Tulip wood is white soft and suitable for carpentry work and also prove useful as a source of paper pulp. The wood shows resistance to fire and used for blacksmiths bellows. Seeds are eaten after boiling. Bark powder is used in skin diseases. The bark decoction is given in dysentery and gastronitestinal disorders. The green leaf infusion is by tribals to poison the wild animals. African hunters used decoction of the fruit for killing the birds.

Place in bioaesthatic planning

Tulip is a very handsome attractive when grown as groves. The tree is useful as shade given tree and can be grown as avenues in city, town and metropolis. Tulip tree is indispensable is parks, stadiums, Front yards of star hotels and horticultural studies for beautifying the surroundings.

6.23 Scarlet Bottle Brush tree (*Callistemon lanceolatus* Sweet)

Family: Myrtaceae

Vernacular names

Hindi: *Lal botal brush*; Telugu: *Seesa brush*

Geographical diversity

Bottle brush tree is introduced from Australia. It is well distributed in North India particularly in Uttar Pradesh, Punjab and Haryana.

Morphological features

Bottle brush is very handsome evergreen tree with a group of leaves at the end of small branches willow. The flowers are scarlet in colour and floral tufts or inflorescence appears like bottle-brush. The massed red coloured inflorescence gives good beauty to the tree. The stamens and filaments project in all directions near the tip of branches. These appear like bristles of brush. Its growth is somewhat interesting. At the end of flowering season, the stamens dropped to the ground and pods arise from the bases. The pods have five seeds. The pods dehisces on the tree and seeds dropped to the ground and germinate on swampy places. The flowering branches continues to grow beyond the point and bears next bloom. The seed pods will hang at the end of branches upto one year. The flowers display their beauty at the and of winter.

Economic utility

The wood can be used the making boxes, tool handle sand other agril implements.

Place in bioaesthatic planning

The trees can be planted in the middle and corners of lawns, road sides and compounds of hospitals and offices.

6.24 Noble Amherstia (*Amherstia nobilis* wall)

Family: Caesalpiniodesae

Vernacular names

Burmese: *Thawka*; English: *Splendid Amherstia, Flame Amherstia*; Hindi: *Amherstia*

Geographical diversity

Amherstia is a native of Myanmar and widely formed in West Bengal, Assam, Bangladesh and Kerala where humid climate prevails. Very good speciments can be seen in the Gardens of Kerala and West Bengal.

Fig. 6.5: Noble Amherstia (*Amherstia nobilis*), Bark – grey, L – pale copper/purple hue, Fl – vermilion ( ) golden yellow spots, Fr – brown pod.

Morphological features

Amherstia is a medium sized evergreen tree with compound leaves. The leaves are droopy and paripinnate compound. Young leaves are pendent, flaccid and appears charmingly. The young leaves are pale coppery or purple hue and later turned to green. The stem is stout trunk with ashy grey colour. Flowers are large and petals contains sprays of vermilion red stripped with golden yellow spots. They are arranged in long, pendent candlabra – like racemes borne in the axils of leaves. The tree displays its blooms during November to April. The flowers last its life within three days. The fruit is pod, flat shaped with beak at the end.

Economic utility

It is an excellent tree for planting in gardens wood is used for making plants and packing cases.

Place in bioaesthatic planning

The tree flowrish in warm areas of South India and damage may occur in North India due to frost and hot dry winds. The tree can be planted as avenue on the side of roads, Airports, Railway station and public parks.

6.25 Lahura (*Tecoma undulata* G. Don)

Family: Bignoniaceae

Vernacular names

Hindi: *Rugtrora*; Marati: *Rakhtreora, rakhtroda; Roira*; Mumbai: *Lahari, Lehero*; Punjabi: *Luar, Rohira*; Sanskrit: *Rohi*

Geographical diversity

Lahura is mostly formed in subtropical regions. It is very common in arid areas and abundantly formed in desert areas of Haryana, Rajasthan and Gujarat.

Morphological features

Lahura is a small tree and highly attractive when in full flowering. The branches are droopy and foliage is grayish – green. It has conspicuous wavy edges. The flowers are tubular – shaped and bright orange yellow. About 5-10 bunchy flowers borne at the end of small lateral branches. They come to flowering from march to April and highly handsome. The growth of tree is very slow. The fruit is curved little, 20cm long and 15 cm broad. The seeds have narrow wings.

Fig. 6.6: Lahura (*Tecoma undulata*), L – greyish green, Fl – orange yellow, Fr – brown.

Economic utility

The wood being hard and is used for carvings, domestic furniture, and agricultural implements. The wood is prized for furniture and fire resistant and attack of fungus. Leaves are used for cattle fodder and relished by goats. It is highly useful for afforestation and to prevent land sliding in dry tracts.

Place in bioaesthatic planning

Lahura can be planted as shrubbery in public parks and home gardens. It can be planted as an avenue tree on road sides, approach of aerodromes, Railway stations and public offers. They display bright orange – yellow flower at the end of winter.

Chapter

7

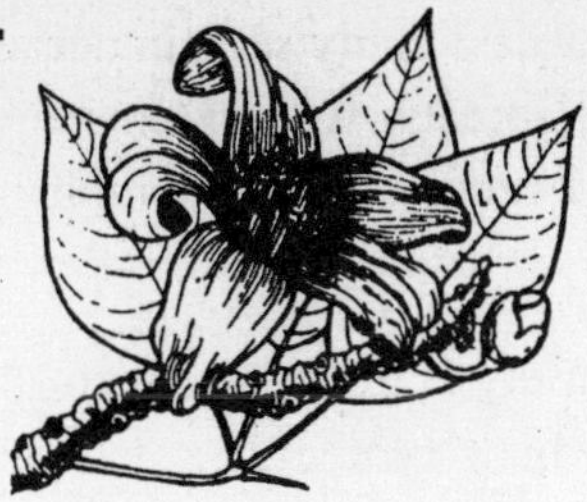

Elegant Trees

"The brilliant Sun sinked toward the West
The night queens and Jasmines burst opened
The delicious floral scents filled the air
Waiting for the tender touch of Moon's rays"

7.1 Champa (*Michelia champaca* L)

Family: Magnoliaceae

Vernacular names

Gujarati: *Rae-champo;* Hindi: *Champa,Champa*; Malayalam: *Champagam*; Marati: *Kud-champa;* Oriya: *Kanchanam*; Punjabi: *Champa;* Sanskrit: *Champaka;* Tamil: *Shembuga*; Telugu: *Champakam*

Geographical diversity

The tree is widespread in all countries of Tropical Asia. It is wild in the forests of eastern sub-Himalayan zone from Nepal eastwards, along the foot-hills upto Assam. It also found in the Western Ghats from

Kanara to Kerala. Champa is occasionally seen in deep valleys cooled by perennial springs in Bihar forests. It is extensively cultivated in almost all parts of India and Myanmar, around temples for the sake of fragrant flowers.

It is capable of growing on a variety of geological formations i.e. gnesis, slates, Quartzites and schists, which weather to a rich sandy loams. The tree grows luxuriously on moist, deep, well-drained, good quality soil. It is very fast growing, thriving best in damp situations on moist, deep and fertile soil.

Morphological features

Champa is a tall, handsome, evergreen tree, with a close, tapering crown composed of ascending branches, attaining a height of 35m with a girth of 3.5-4.0m. In its habitat, it surpasses all its association in its excellent form. The bark is ashygrey and rough outside. The leaves lanceolate, ovate, acuminate and glabrous. Flowers axillary, pale yellow and very fragrant. Fruits consists of long clusters of capsules on a spike. Seeds are brown, polished, hanging out on long cords. The tree flowers early in its life. The new leaves are produced about March in yellow flowered variety but white flowered variety mostly appear later the hot season. The blooms possess such a strong scent that even the bees are said to find the perfume too heavy and to pass them by. Parrots and other birds devour large quantity of seeds when they ripe and reduce the natural reproduction.

Economic utility

The timber is lustrous, smooth and heart wood is olive-brown. It is light, soft medium textured with distinct growth rings. Timber is easy to saw, finishes to a smooth, glossy surface and polishes well. Excellent veneers can be made with wood. Champa timber is excellent for making light furniture and indoor works. It is highly suitable for heavy packing cases, battery separators, pencils and superior type boxes. As it is considered to be a Grade I wood for moisture-proof, the wood is used for ship, boat building, cabinet-making, general joinery, carriages, ploughs, toys and door panels. Excellent drums are made in the West Coast with the wood. Camphor is extracted from the wood by distillation. All *Michelia* spp can be a good substitute for *Sitka spruce* used for air craft work.

The flowers are used in religious ceremonies and are worn by South Indian ladies in the hair. Fresh flowers yield a fragrant champaca oil which is employed in Jasmine flower perfumes. The perfume '*Sampanghi*' oil is also prepared from Champaka. The leaves, roots, bark and fruits are medicinally and used in Country medicine.

Place in bioeasthatic planning

Champa trees can be planted as avenues in national highways, around parks and on hills near the towns and cities.

7.2 Coral Jasmine or Night Jamine (*Nyctanthes arobor-tristis* L)

Family: Oleaceae

Vernacular names:

Bengali: *Sephalika, seoli*; Gujarati: *Jayaparvati*; Hindi: *Harsinghar, Seoli*; Kannada: *Harsing*; Malayalam: *Pavizhamalli*; Marati: *Khurasli, Parijatak*; Mundari: *Saparomum Kulamarsal*; Oriya :*Godokodiko*; Sanskrit: *Parijata, sephali, a*; Tamil: *Manjhapu*; Telugu: *Kapilanagadustu, Pagada malle, Parijatamu*

Folklore

People who walks nearby coral Jamine grove during rainy season may feel the excellent smell which delight both heart and soul. One can understand why the coral Jasmine called as *Rat-Ki-Rani* or *Queen of night*. When the fragrant breezes go through his nasal openings. Surprizingly the flowers open only during dusk and emits most pleasant smell throughout the night and fall-off at the dawn of next day. The scientific name "*Nyctanthus*" is derived from greek word which word which means '*night flower*'. The species name '*arbor-tristis*' means '*sad tree*' It looks like sad during daytime by closing its flower. People fear to sit near the tree because they believe that jasmine scent attracts the cobras and other serpants.

According to *Bhagavata*, Satyabhama, the dearest wife of Lord Srikrishna asked her husband to plant the 'parijata' in her garden to enjoy its fine fragrance. Lord Krishna fought with Indra, defeated him and brought down the night jasmine from heaven. The story of '*Parijatapharan*' (Stolen parijata) was described by several sanskrit poets in their *Kavyas*.

According to '*Vishnu purana*' once parijataka is a daughter of a king, fell in love with Sun God. But Sun God deserted her. With that angry, Parijataka killed herself and was burnt. A shrub was borne from her ashes. It can not bear the looks of Sun and drops its flowers in the morning. The plant is called as '*sad tree*' from that day onwards.

Geographical diversity

The tree grows all over India from sub-Himalayas to Cape comerin. The tree is very hardy and can withstand any type of soil. However, heavy soils with impeded drainage is not suitable.

Morphological features

Coral Jasmine is a large shrub and some times it was found as small tree. It can not grow beyond 4m height. The leaves are opposite and arranged in pairs. The leaves are rough. The flowers are borne from axils of leaves and arranged in small terminal cymes. The flowers are white with a ring of orange colour. The perfume of the flower is like jasmine or *Cestrum nocturnum* which helps in pollination.

Economic utility

The bark used for tanning, corolla tubes were used for dyeing silk in ancient times. Corolla tube contain a colouring matter, *nyctanthin*, which is identical with crocetin from saffron. Fragrant flowers are offered in temples and made it into garlands. Scents are prepared from the flowers. Seeds yield a fixed oil. Leaves are expectorant, antibilious and used in the treatment of rheumatism, and feavers. Decoction was used in sciatica. Juice is anthelmentic, expectorant and Cholagogue. Powdered seeds are used for scurf of scalp on the head.

Place in bioaesthatic planning

These Shrubs can be planted in gardens, Front yards of houses and in public parks for the perfume of flowers.

7.3 Devil's tree (*Alstonia scholares* R.Br)

Family: Apocynaceae

Vernacular names

Bengali: *Chattim*; Hindi: *Chatian, Chhatwan, Satni*; Marati: *Satvin*; Sanskrit: *Saptaparni*; Tamil: *Palaigh*; Telugu: *Palakarra, Palagaruda*

Geographical diversity

Devil's tree in evergreen and dense stands can be seen in Western and Eastern Ghats upto an altitude of 1000m. It is not very Common in North India, but has been met with on the banks of Himalayan streams and brooks in Rishikesh and Kalsi. The tree prefers high rainfall areas and moisture regions. The specific name, *scholaris*, refers to the fact that the wood is mostly used to prepare slates which are used by school going children. The wood is also specially used for making blackboards used in high schools and colleges.

Morphological features

Devil's tree is a tall evergreen tree with a grey bark and verticillate branches. The stem often flutted or buttressed. The branches are very much like in ribs of umbrella. Leaves 4-7 in a whorl, leathery, shape like mango tree, dark green above, whitish bloom beneath. The leaves are obovate, eliptic oblong, with numerous lateral nerves. Inflorescence is compact, umbellately-corymbose cymes and the umbels whorled. Flowers are white or greenish white and sweetly scented. Fruit is a long follicle, pendulous in pairs, produced in large numbers like hanging tresses from a head of young lady. Seeds slender, flattened, peltately attached, densely ciliate with long hairs all round. The tree flowers from December to March and fruiting occurs in the months of May and June.

Economic utility

The wood is whitish, even-grain, porous, soft light but seasons badly. The wood is not suitable for permanent structures. It is used for making tea-boxes, packing cases, black boards, slates and photoframes. The wood is widely used for making coffins and scabbards. The *Dita bark* employed as a bitter and astringent in the treatment of dysentery and fevers.

In the rural areas, the tree is called *Shaitan* as the people believe that the tree has maleficent magic properties. Tribals of Western Ghats are not willing to sleep or sit under the tree because of belief that it is a sure token of death by devils which guards the tree. Inhabitants of Eastern Ghats are unwilling to pass under the shade of tree while walking due to fear of evil spirits. The tree is considered as abode of demons and eerie spirits in Western India and unwilling to see in the early morning.

Leaves used in beri-beri, dropsy and congested liver. Latex applied to sores, ulcers, tumours and rheumatic swellings. Flowers yield an essential oil which acts as a depressant on central nervous system. Ash of the plant employed as caustic to open abscesses. Wood charcoal is used as gun-powder.

Place in bioaesthatic planning

The tree is very elegant because of very regular branching and masses of flowers that envelope the tree outside. It can be planted along the roadsides of towns and metropolitan cities.

7.4 Indian Cork or Jasmine tree (*Millinotonia hortensis* Linn. f)

Family: Bignoniaceae

Vernacular names

Bengali: *Cork-gach*; Hindi – *Akas neem, neem Chameli*; Kannada: beratu; Kat: *malli, Karkku.*; mach: *mach,sitabara,bakeni*; Malayalam: *Katesam*; Marati: *Kavla nimb, akas nimb*; Mini: *chambeli*; Oriya: *reali*; Tamil: *mara-malli*; Telugu: *kavuki*

Geographical diversity

The tree is widely adopted to various climatic conditions and can be grown throughout India. It's rank growth can be seen in moist regions where rainfall is abundant but also perform well in dry climates. It is very hardy to variations of climatic parameters but susceptible to damage during cyclones and storms.

Morphological features

Jasmine tree is a fine tall, handsome straight ever green tree. It is having narrow crown and grows to a height of more than 25m. The stem is covered with Yellowish grey corky bark. The branches are droopy and consists of deep-green coloured leaves. The compound leaves have many leaflets usually 3 4 pairs. The tree usually shed its leaves during December and January. New leaves are born from January to March when temperature gradually rises.

The inflorescence is terminal panicle and consists of several silvery-White small flowers. The flowers emits sweet scent during nights. Flowering occurs from November to December and later from April to June. The ground is full of white carpet during flowering season as the small flowers drops from the plant. People believe that these fragrant flowers attract the snakes and do not prefer to grow in the house compounds.

The fruit is small, slender, flattened pod, having points at both ends. The trees do not fruit normally in India but occationally thin and flat seeds are formed with delicate wings.

Economic utility

The wood is useful for furniture and other ornamental work. Cork is extracted from the bark. The bark contains a bitter principle which is used as antipyretic. The wood is used for the manufacture of tea-boxes, drawing boards and plane tables.

Place in bioaesthatic planning

The trees can be grown as ornamental tree in the front yards of houses. It can also be used as excellent avenue tree in both rural and urban areas. The plant can be propagated through suckers and seeds. In coastal areas, the tree damage during strong winds due to brittleness of branches but recuperate with rapid growth.

7.5 Siris (*Albizzia lebbek,* Benth)

Family: Mimosaeae

Vernacular names

Bengali: *Sirish;* Gujarati: *Pilo sirshio*; Hindi: *Sirs, Sirsa;* Kannada: *Begemara*; Malayalam: *Vaga*; Sanskrit and Marati: *Sirisha*; Tamil: *Vegei*; Telugu: *Dirasana*; Trade-*Kokko*

Geographical diversity

Siris is a large deciduous tree and mostly found along the banks of streams rivers, and brooks, It grows from plains to high altitude areas and needs high water requirement. New leaves appear during March-April and flowering occurs in hot summers (May). Fruit formation takes place during October-November but remains long on the tree.

Morphological features

It has a spreading crown with large trunk. The bark is dark-greyish in colour and irregular cracks are found. Leaf rachis 10-18cm long, stout with a large oblong gland near the base. Pinnae has 4-9 pairs of leaflets which are oblong, obtuse, and glabrous with short petiole. Inflorescence is short corymbose receme with many flowered heads. Flowers are greenish White, fragrant with short pedicel. The pods are flat thin, firm and straw coloured. Each pod contains 6-10 seeds which have brown spots.

Economic utility

The tree is often planted in avenues. The sapwood is large and white while the heart wood is dark brown, mottled and streaked with dark colour and used for making house furniture. The wood is very hard, durable and seasons well. It takes fine polish and can be employed for making dining tables, wooden chairs and dressing tables. Being very hard nature, it is employed for making sugarcane crushers, oil extractors, well-curbs, wheels and other hard furniture. Camels eat the leaves and twigs voraciously. Tree yields a gum which is used in book binding and paper envelope industry. Bark contains about 8-11% tannin and used for tanning fishing nets. Bark is used for dressing boils. Leaves and seeds are employed in Indian medicine to cure eye troubles.

Place in bioaesthatic planning

The trees can be planted along with national highways, broadway roads, and along the canals. The tree is a rapid grower and can be easily propagated by cuttings.

7.6 Temple tree (*Plumeria ecuminata* Ait)

Family: Apocynaceae

Vernacular names

Assame: *Gulanchi*; Bengali: *Dalanphul, gorur champa*; English: *Pegoda tree, Frangipani, Champa*; Gujarati: *Rhada champo*; Hindi: *Chameli,gul-e-chin*; Kannada: *Devaganigile*; Malayalam: *Ezha Champakan, arali,*; Marati: *Khair champa, Son Champa*; Sanskrit: *Kshira champa*; *Sontal: Champa Pungar*; Tamil: *Arali, perungali*; Telugu: *Nuru verahalu, arhataganneru*

Geographical diversity

The tree is native of Mexico and Guatemala but extensively adopted all over the country. The botanical name *"Plumeria"* is named after a French botanist, *Charles plumier*, acuminata refers to tapering leaves. It is well adopted to both in tropical and subtropical regions.

Morphological features

Temple tree is a small deciduous tree with crooked stem. The white stem and branches appear like lime-washed. The wood is soft and bark produces a milky juice when injured. The leaves are large, broadly lanceolate, tapering at both ends. The venation is parallel and appears distinctly. Leaves are borne in crowed spirals at the end of long branches. The tree is leafless during winter (December onwards) and continuous upto the break of monsoon. New leaves borne at the start of rainy season. The young trees do not shed their leaves and remain highly attractive throughout the year.

The flowers are waxy white with yellow centres. They are highly fragrant and borne in clusters upwardly. The petals are tinged with pink below. The tree flowers during March to April and again during the rainy season. The sweet scented flowers are said to be highly affectionate to *Maha Rudra*.

That is why, every temple in Godavari deltas are having these trees for worshipping the God. Fruits are two horn like pods on a long pendent stalk and seeds are covered with a tassel of wool. Seeds are not produced under Indian climate.

Fig. 7.1: Temple tree (*Plumeria acutifolia*), L – dark green, Fl – white with yellow colour

Economic utility

The tree is much valued for ornamental planting in Hindu temples and public parks. The bark is used for relieving fevers, heal sores and tumours. Bark is stimulant and its decoction is used as a purgative, febrifuge and emmenagogue. The decoction of bark is useful in venereal affections, and said to be a good antiherpatic. The latex is rubefacient and applied in itch, gum trouble and rheumatism. Root is cathartic. Wood is free from termites and used for making drums.

Place in bioaesthateic planning

The beautiful trees are highly suitable for planting in public parks, middle of lawns, factory premises and in college roads. It grows and blooms even in stony soil or even roots are exposed completely. It thrives well even in harsh environments.

Chapter

8

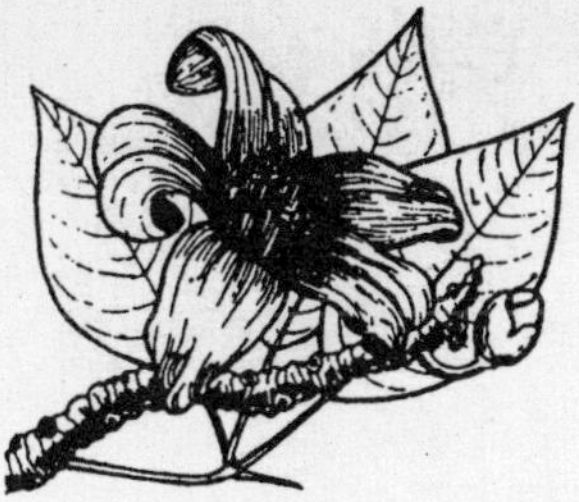

Strange Looking Trees

"Woodman, spare the tree
Touch not a single bough,
In youth it sheltered me,
And I' will protect it now"

8.1 Baobab or Monkey – bread tree (*Adansoni digitata* L)

Family: Malvaceae

Vernacular name

Africa: *Calabash, Baobab;* English: creamof tartar; Gujarati: *Gorakh Chinch;* Hindi: *Gorakh imli, Goram lichora;* Malayalam: *Perauka, Paparapulia*; Marati: Choyarichich; Kannada: Magimavu; Tamil: *Perruka*; Telugu: *Bana Potta Chettu, Brahmarakshsi Chettu*

Fig 8.1: Baobab (*Adamsonia digitata*), L – green, Fl – creamy white, Fr – white green

Popular lore

Baobab is a strange looking tree and it is about the thickest in the world, grotesquely fat, suddently tapering into thick branches. The tree is indegenous of tropical Africa and the trunk reaches 10m. diameter or more. Some trees can be seen even with more than 30m. in girth. *In other words, it is enough to say that about 16 tall men should spread their hands to encircle trunk of the tree.* Another strange feature of the tree is the venerable old age and live about 4-5 millennia under favourable conditions. A French botanist, M.Adanson was noticed some old trees of about more than 5, 300 years old in African forests. The carved letters of visitors in the 14th century can be seen even now on these old trees. The perennial life of these trees can be compared with Gaint Redwood trees of California and coniferous trees of western U.S.A.

The tribals hollowed out these big trees by cuttings the pith and made into living houses. In some of the excavated trunk houses, nearly 30 men can lie down and sleep. Owing its peculiar appearance, most of the rural people imagine the tree as *eerie* and *sinister*. It looks like *vetala* when one look in dark nights. One can remember the stories of tales of "*King Vikram and the Vetala*". In stormy night when gusts of wind shook the tree, the sound is like moaning of Jackals and eerie laughter of spirits. The superstitious awe might also be due to the fact that theses plants are grown in cremation grounds in Africa. The tribals hang the corpses of witchmen, kings and other distinguished leaders inside the hollowed trunks and close them like mummies of Egyptian pyramids. Such bodies are said to be dehydrated and mummified without any chemical treatment. That is why, the tree is considered as demon tree by African tribals.

It is believed that *Guru Gorakhnadh* preached to his desciples under the tree and it was called *Gorakhimli* in Gujarat.

Geographical diversity

Baobab tree is native of tropical Africa and requires warm climate for its growth. Muslim traders introduced into India several centuries back. The tree is now found scattered all over India particularly along the Western coast. It was also found in U.P, A.P, West Bengal and Tamilnadu particularly in Cities.

Morphological features

Baobab is an erect tree grows upto 20m high with a girth of 10m. The bark is very smooth, occationally scarred and unsighty. The leaves are large and resembles that of silk cotton tree with a long stout petiole. About 5-7 leaflets are borne from the tip at the same place. The flowers are large, white with a touch of purple in the middle and showy. The fruits hang down from the stalk, greyish or pale brown in colour. The fruit looks like small gourd. Numerous seeds are embedded in the woody covered fruit. The pulp is eaten by tribals. It refreshes in the hot summer. Monkeys are very fond of the fruit and eat with relish.

Economic utility

The soft nature of its wood makes the tree useless for carpentry work of several types but it is highly suitable for the match industry. The wood is employed for making trays, canoes, rafts, floats and fishing-nets. The grotesque trunk sometimes becomes hollow and forms water reservoir: Tribals sometimes store about 5, 000 litres of water during monsoon season and used in off-season. It is said to be one of the most useful tree of African continent and every part yield some kind of economic product. The bark yields a soft fibre but very strong. The fibre is used for the manufacture of ropes, bags, sacking, clothing and strong which eminently suitable for printing currency.

The fruit pulp tastes like cream of tartar and used as refrigerant in various fevers. It is used in the treatment of diarrhoea and other stomach disorders. A pleasant and cooling *sherbat* is prepared from the fruit pulp is also diaphoretic and gives relief in bronchial asthama and allergic dermatitis. Tender leaves are used for seasoning. The dried leaves are credited with increasing perspiration and preventing kidney troubles. The powdered leaves when orally administered to guinea pigs prevents the crisis in asthma induced by histamine aerosols. The Kernels are eaten as nuts and yield an oil which is used for cooking.

The ash of the fruit and bark boiled in oil and used as soap by tribals. The dry fruits are used as net floats by fishermen of West Coast. The bark is exported to Europe for extracting drug which is used for malaria treatment in place of *Guinine*.

Place in bioaesthatic planning

It is not suitable for planting along streets or avenues on account of large trunk. As it is an interesting tree, it can be planted in public parks, college and university campuses, and hospital premises. In open places it makes a brave show of blooming.

8.2 Cannon-Ball (*Couroupita guianensis*)

Family: Myrtaceae

Vernacular names

Hindi: *Topegola;* Telugu – *Talakalyala Chettu, Volley ball chettu*

Geographical distribution

Cannon-ball is a native of South America and introduced into India and Sri Lanka duirng the British rule. It is grown vigorously in warm and moist areas where rainfall is high. Beautiful trees can be seen in public looks very strange because of its untidy branches and large rounded fruits. The fruits are about the size of human head. That is why it is called as *cannon-ball tree.* It can not tolerate temperate and arid Climates and flourishes only in moist warm regions.

Morphological features

The tree grows to a height of more than 10 metres. It has some strange morphological feature like untidy branches and hanging flowers from the trunk. The leaves are dark green, borne at the tip of branches like clusters. The flowers borne from the main stem and hanging downwards. The flowers occupy from Crown to ground. It is very curious and attractive to look at the tree during flowering because reddish flowers are very large, fleshy, Pink and white with yellowish colour on the outside. The fruit is globular pod, brownish in colour and grows to a size of adult man's head. The fruit takes about one year for complete ripening. The seeds are embeded in stinking fruit pulp. It is propagated mainly from ripened seed.

Another strange feature of the tree is the habit of dropping all the leaves at once many times during the year. New leaves arise rapidly within one week or 10 days and covers the tree. The hanging flowers like rat traps and large fruits about the size of cannon ball are highly curious to the viewers.

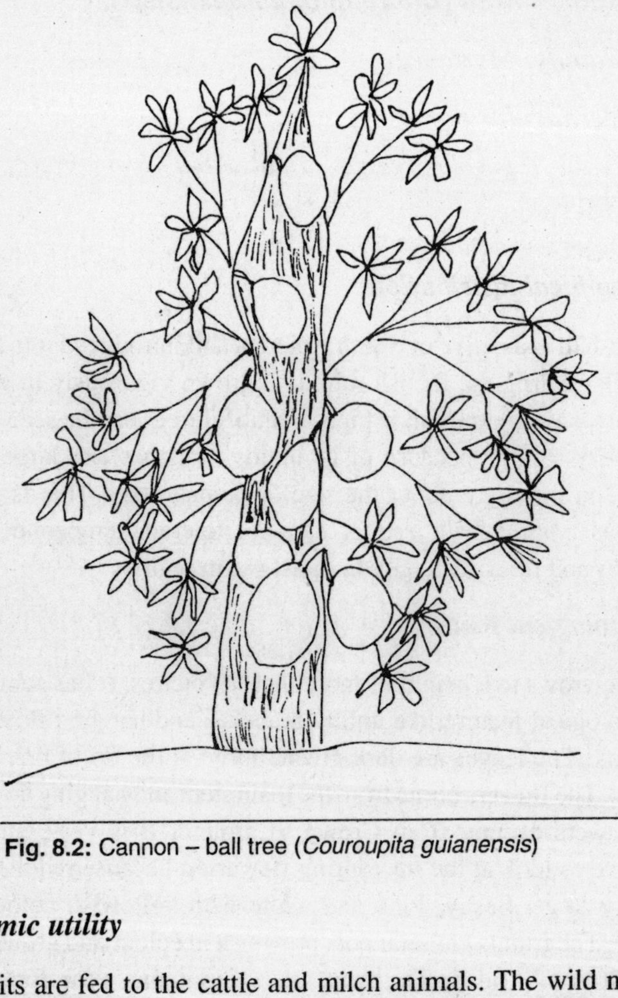

Fig. 8.2: Cannon – ball tree (*Couroupita guianensis*)

Economic utility

The fruits are fed to the cattle and milch animals. The wild monkeys relish the fruits and eaten voraciously. The inhabitants of Guiana feed them to domestic animals as supplementary feed.

Place in bioaesthatic planning

The tree can be planted in public garden, parks, along the town roads, and around the lawns of star hotels.

8.3 Fish-tail Palm (*Caryota urens,* Lim)

Family: Palmae, Arecaceae

Vernacular names

Assam: *Baraflawar, berlimad, bherawa, surmadi*; Gujarati: *Shankarjata, Shiva Jata*; Hindi: *Mari*; Kannada: *Bagani*; Malayalam: *Anapana, Kundapana, vazapana;* Marati: *Berli*; Oriya:*Salopa*; Sanskrit: *Mada, dirgha*; Tamil: *Tippili, Koondalpanai*; Telugu: *Jeelugu Chettu*

Geographical diversity

Fish-tail palm or Indian sago is indigenous to India. It is distributed from Sub-Himalayan regions to cape comerin. These wild palms are Quite Common in the forests of Western Ghats where the rainfall is heavy. This tree was first introduced into Botanic Garden of Calcutta during the British period. It also found in the forests of East Coast, Bihar, Orissa and Northeast Indian states upto 2000m elevation from the sea-leavel.

Morphological features

The palm is 15-20m tall but on occasion it may attain to a height of 30m with a diameter of over 30cm. The trunk is grey and smooth. The leaves are 5-7m long with 10-20cm long leaflets which is largest of any Indian palm. They are bipinnate and leaflets are like a fish tail and Wedge shaped. It contains strong parallel nerves with serrated tip. Spadix is 3-4m long, with dripping branches. Fruits are drupaceous contain 1-2 seeds. These are wrinkled with grooves and furrows. Floweing occurs in hot summers and fruits are formed in August when rainfall is at peak.

Economic utility

Many products are obtained from the plant and every part has its economic value. The palm oozes plentiful of sweet juice from young spadices. This *neera* is much relishing and taken by local people as a cool drink. A fully grown 25years old palm yields about 15 litres of juice within one day. This soft drink is highly nutritious and taken by tribals. Toddy is formed after fermentation of sugars but intoxicating. Fermented toddy is much relished by local tribals and drink this juice continuously without attending their agricultural works. Arrack and Ethyl alcohol can be made after distilling the fermented Juice. The fresh juice is boiled and made into jaggery after solidification. This is called Palm sugar.

The tender folded leaves in the apical bud is used as vegetable in Western India by rural poor. The pith of palm consists of high quality sago which is known s *bastard sago* in foreign markets. It was almost equally good from *Metroxylon sago* which is considered as best sago. Tribals of Gujarat, Rajasthan and Punjab mix the flour with millets like Pearl millet and great millet for making *Chapaties.*

Place in bioaesthatic planning

Fish tail palms can be planted on side of ways in public gardens and University Campuses.

8.4. Sausage tree (*Kigelia Pinnata* D.C.)

Family: Bignoniaceae

Vernacular names

Hindi: *Jhar phanoos;* Telugu – *SoraKaya Chetta*

Geographical diversity

The sausage tree is the inhabitant of tropics and its native home is Mozambique of West Africa. It is highly adaptable to climatic variations. Though it prefers warm regions it performs equally in cold climatic areas like Punjab and Haryana. Under fertile and well-drained soil it grows well even in arid and semi-arid tropics. However, luxurious growth was found in humid tropics where rainfall is high.

Morphological features

The tree is evergreen and medium in size with dark green foliage. Under favourable conditions, it attains to a fairly large size. The tree is very peculiar to see at the time of flowering and fruiting. The scientific name *Kigelia* (genus) derived from its native name of Africa and *Pinnata* (species) refers to the *"pinna shaped leaves"*. The tree has many branches and arranged in dense manner. The canopy is like a round head. The leaves are pinnately compound, rough dark green and arises at the tip of branchlets.

The inflorescence is long hanging raceme. The flowers are large, appears like cups and mottled dark purplish in colour. The flowers emits foetid smell. The floral inflorescence appears like hanging candlebars. The tree is at full bloom during summers (April to May). The flowers are cross pollinated by the bats. The fruits are large gourd like with heavy weight. These sausage like fruits hangs from long cord like stalks and occationally 2-3 together.

Economic utility

The tree can be grown in public places where dense shade is needed. Fruits are used as a purgative in rural areas. The fruit pulp is employed for dressing syphilitic sores. Bark extract is used in dysentery, rheumatism and veneral diseases. The seeds are roasted and eaten by tribal people at the time of famines.

Place in bioaesthatic planning

As the tree is having attractive foliage and peculiar fruits, it is well suited to plant in the public parks and amusement places. The foliage is also having some repellent smell, cattle do not prefer to feed the branches. The plants need not be protected against cattle grazing. The tree is most commonly propagated through seed and layering.

Chapter

9

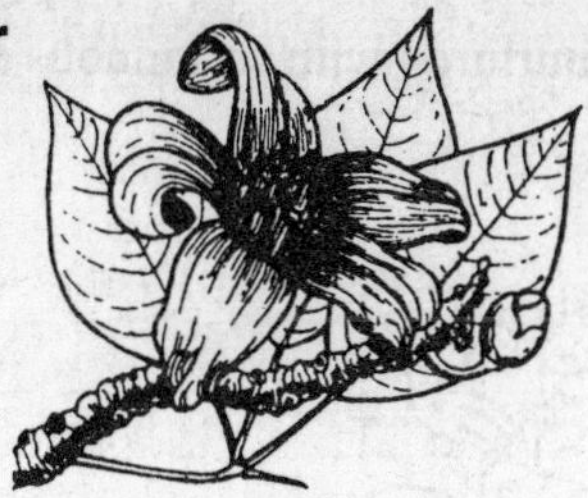

Bonsai: Miniature Trees

"Bonsai is exquisite creation of myriad moods of trees through patient perseverence with a love of beauty".

The word '*bonsai*' (pronounced bone-sigh) is derived from Japanese word '*bon*' means a shallow container, and '*Sai*', meaning fine art like painting and sculpture. The symmetry creates a feeling of beauty like a picaso's fine painting. Truly speaking the art of bonsai was born in China around 200 BC and referred as '*Pan-tsai*' or '*Pan-Ching*' at that time. Later the art spread into Japan, flourished there and came as a noble art and emanated world over. It was refering as japanese art like a an average American, despite his European, or Levantine or African or Chinese or Indian. The Japanese have spent centuries for developing the dwarfing of plants until it has achieved the status of an art. Unlike the sculptor's hard rock or painter's or muscian's instrument, the bonsai artist's medium is living thing growing continuously with soul of its own. Understanding its nature and patient perseverence is needed to get desirable results. One has to remember that *Rome was not built in a day*.

A good bonsai is a love of beauty nurtured with continuous efforts and unbounded patience.

Fig 9.1: The bonsai that displaying its charming single flower with dwarf stem.

9.1 Dwarfing the plants – sizes

Dwarfing the plants by certain cultural practices like planting in shallow containers which have little space for root growth resulting in limited shoot growth. In addition to restricted root system, Judicious pruning of plant parts above ground make the plants dwarfed.

There are 4 types or sizes of bonsai (according to height)

a) Mame bonsai (Bean size bonsai) – 5 cm

b) Small bonsai – less than 10 cm

c) Medium bonsai – less than 15cm

d) Large bonsai – 30-60cm

Based on container size

a) *Smallest* – The containers can be carried in the palm of hand

b) *Middle size* – The containers can be transported through using two hands

c) *Largest size* – The containers need two persons to change the plants from one place to another.

9.2 Planting in the Pot

The planting spot is mainly depends upon shape of the container. If the container is rectangular shaped, the plant should be placed at one end preferably at a point seven-tenths of the distance from one side. The plant should not put in the middle of container which is not good from an aesthatic point of view. The training of the growing plant is easy when it planted on one side.

Before putting the plant in the container, the roots should be checked and excess roots and old roots may be cutted to make the root system small. Press the soil in and around the roots after planting to expell the air cells inside. Watering should be done immediately as soon as the plant is potted.

The pot should be kept in half shade upto 10 days for quick recovery. The foliage may be sprayed with water frequently to avoid wilting due to water stress. After one week, the pot may be kept in open sun, in the forenoon, for about 5 days for carrying the process of photosynthesis. Later, the plants can be exposed to bright sun for a full day.

9.3 Suitable soil

All soils are not suitable for growing bonsai plants. One has to understand the soil properties, its nutrient supplying capacity and mineral requirements of plants that are selected for growing as miniatures. In fact, soil is not merely a group of mineral particles but also has biological system of living organisms as well as some other components. The soil air is mainly depends upon size of soil particles. Soil hold the gases or air in between the space present among soil particles. Dry soil rarely has CO_2 content of more than 0.5 – 1% or an O_2 content of less than 20%. The wet soil contains CO_2 and O_2 in equal proportion. The colour of the soils also varies greatly from region to region. The colour also indicates the type of soil and its mineral compounds. In temperate regions, the colour is usually dark owing to presence of high organic matter. The red and yellow soils show the presence of iron oxides where as greyish soils have quartz, carbonate and various ferrous iron compounds. The grower of *bonsai* should modify his soil either by mixing compost or coarse particles, according to their texture.

9.4 Characteristics of good soil for bonsai

- It must supply all the essential nutrients, soil water and oxygen for atleast a year.
- The texture should be such that surplus water should be drained immediately.
- It should able to absorb and retain sufficient moisture for the constant supply to the plant.
- Oxygen supply should reach even to fine roots to avoid rotting.

9.5 Improvisation of heavy soils

If the soil contains high clay content, it will lead to ill drainage and unhealthy environment in the soil system. Sand may be added to improve the drainage conditions. Coarse sand which is used by builders is preferable than fine sand obtained from river beds. Sufficient well decomposed compost may be incorporated into the soil to reduce the swell-shrink property of soil.

9.6 Tackling of sandy soils

Being light textured, sandy soils are poor in nutrients and less water holding capacity and can not able to sustain plant growth for a long time. These coarse textured soils have a insufficient humus content that play a great role in the retention of nutrients and soil moisture. Sufficient quantities of compost or leaf mould may be added to the soil. Some times, little quantity of sand is added to the compost. The role of sand is two fold, improving the draingage property of soil as well as providing little hard particles which are aided in the proliferation of fine roots. When the fine rootlet while growing encounters a hard sandy particle, it will divide into two rootlets in order to avoid the obstacle and thus produce large number of fine rootlets. Such type of soils improve the absorbing capability of trees on water, nutrients and oxygen from a limited space.

9.7 Ideal ingredients of soil

Loam	–	8 parts
Sand	–	1 part
Bonemeal	–	table spoon
Compost	–	1 part

Sometimes compost may be substituted with cow dung, horse dung, sludge and neem cake etc. The nutrients may be supplied according to the particular need of trees. Long experience and critical look is the main guide to determine the quantum of nutrients.

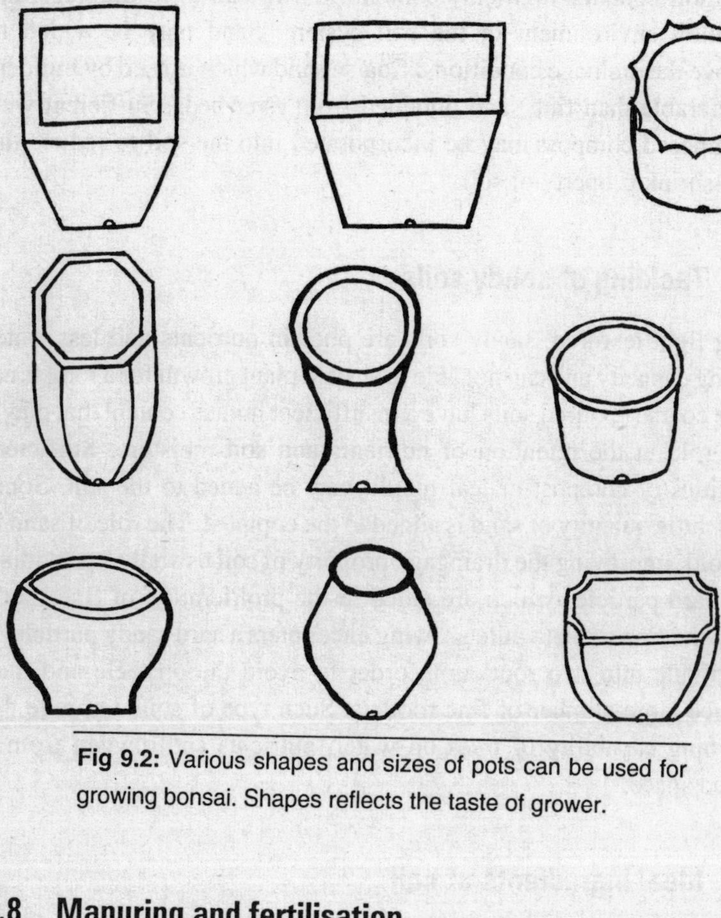

Fig 9.2: Various shapes and sizes of pots can be used for growing bonsai. Shapes reflects the taste of grower.

9.8 Manuring and fertilisation

In order to make the tree a dwarf, neither it should not be starved of nutrients nor thrirstded of water. If the plant is grown in impoverished soil, it may suffer with nutritional disorders become ugly in appearance and die off soon. It is always better to use organic source of manures for better and sustained growth. Plant originated concentrated manures like Sesame cake, cotton cake, Mustard cake, Neem seed cake, soybean cake and Mahua cake are best for supplying all nutrients in a balanced manner; The oil cakes can be supplemented with animal manures such as Chicken droppigs, bone-meal, fish-meal and well decomposed goat pellets. Wood ash and dung cake ashes not only useful as soil conditioner but also supply basic minerals like potassium and prevent the soil from becoming too acid on a long run. Sludge is also good manure and can be used by mixing with well decayed compost. If undecomposed manures are added, there is a problem of offensive odour and these should not be kept indoors for a few days.

9.9 Preparation of liquid manure for bonsai

Ingredients required

Groundnut oil cake or Rape seed oil cake or Soybean oil cake	2 kgs
Wood ash	200 kg
Single super phosphate	150 kg
Plastic bucket (20litre capacity) along with lid	1

The oil cake, wood ash super phosphate can be putted in the bucket and stirred thoroughly to get uniform mixture. The bucket may be filled with water upto 20cm of the top. The entire mixture may be stirred with hand or stick thoroughly and covered with lid. The manure may be allowed to ferment for about one fortnight in hot weather. During the winters, the fermentation process may prolong upto one month due to low temperatures. After complete fermentation, the liquid decanted, filtered and stored in well-corked bottles.

The liquid may be further diluted with water in the proportion of 1:10 or 1: 15 based on the plant requirements before application. Over fertilisation damage the bonsai, results in complete loss. Manuring the bonsai, should be based on years of experience. Under-fertilisation is better, until the grower gain experience of nutritional needs of his trees. The weaker liquid solution may be given during the spring when trees grows to begin vegetatively. There is no need of fertilising during monsoon or rainy season. After the rainy season ceases, the plants may be fertilized with weaker solution before the start of winter. When bonsai going to winter dormant period, it may be fertilized with strong solutions (1:10). Fertilization with strong solutions are necessary during Autumn when flower buds and fruits appear in the leaf axils.

Fig 9.3: Select attractive fruit bearing trees for making bonsai.

9.10 Need of humus in potted soil

The humus is a store house of nutrients. Its gradual decomposition by various microbes results in the liberation of ammonia, carbondioxide and other nutrient ions which are taken by plant for their normal growth. Besides, humus improves the physical, chemical and biological condition of the soil and aids in making the favourable medium for the growth of plants.

Physical

- Humus improves the structure of heavy soils and makes them well aerated. Better structured soils favours root development and decrease soil erosion.
- Humus also improves the sandy soils by increasing their water holding capacity.
- The dark colour developed during organic matter decomposition helps in heat absorption of soil during winters.

Chemical

- Humus supplies nutrients for plant growth in a balanced way.
- Helps in liberation of nutrients from insoluble minerals found in the soil.
- The cation exchange capacity of soil is increased which results in more retention of nutrients in the medium and reduce leaching losses.

Biological

- Serves as a source of nutrients and energy for microbes.
- Metabolism of soil organisms produces cementing material which causes aggregation of soil particles.
- Neutralize the toxic substances and kills harmful organisms in the soil.

9.11 Cleaning and disinfection

The woody bonsai's require cleaning for every fortnight under normal conditions. The trunk and underside of the leaves should be cleaned neatly with fresh water. The trunks should be washed with the help of tooth brush. Otherwise the insects embeded in the bark can not expelled. It is better to spray the plants with mild insecticides of plant origin. Home made sprays are best for disinfection of plants. Decoction of tobacco residues and neem cake can be used as effective insecticide. Butts of cigars and cigarettes can be soaked in water, filtered with a cloth and can be used as spray solution by adding soap to it.

9.12 Pruning of miniatures

Very young trees that are below 5 cm tall should not be cut. The plants of 1-2 year old should be cut-off just above the dormant bud. After one year, the branch may be again cut-off at 2 cm above the base. Such a cutting back for about 5 years results in the development of interesting shaped trunk. The tiny buds that arise from the base may also be encouraged and cut back when they grow fully.

Pruning and trimming are needed for restricting the size and shape of the *bonsai*. Unscientific pruning lead to a grotesque trunk which is unnatural and appears ugly. The best time of this operation is when the tree awakening from dormancy. Remove all unnecessary and dead branches and cut the excessive vegetative growth and long branches.

Some tips to effective pruning

- The branch should be examined carefully for its length and position of bud
- The cut should not be made beyond a twig or bud which is to grow vertically upwards.
- The twigs should not be overcrowded.
- No branch should not cut thoughtlessely across the old wood.
- Pruning process should continue throughout the life of bonsai.
- The leaves should be made into smaller in size and should proportionate to the bonsai tree.

- When new shoots develop, they are allowed to grow until five nodes and pruned back leaving two nodes.
- Pruning should be done preferably as the growth is vigorous as the flow of the sap is speedy.

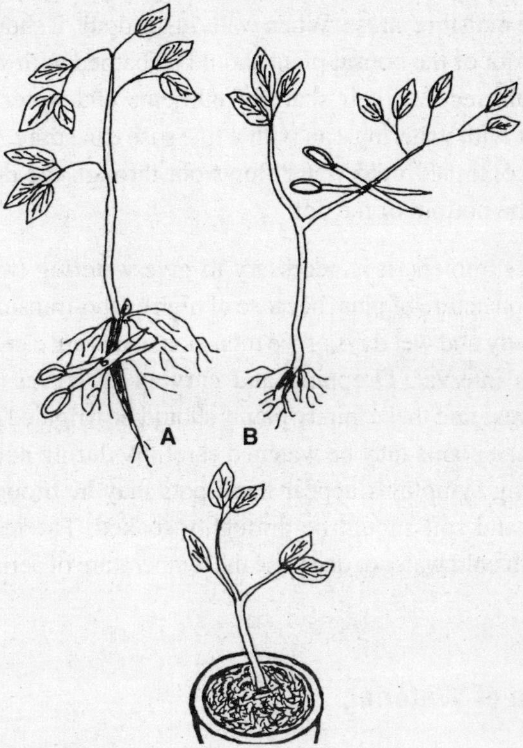

Fig 9.4: Cutting of stem and roots of bonsai periodically (A & B) Reporting of bonsai after trimming (C).

9.13 Technique of watering

Watering is most important for the survival and proper growth of bonsai. Most of the people have a myth of withholding water and nutrients will lead to dwarfing of plants. Such type of belief is quite contradict with actual facts. Truely speaking bonsai requires plenty of water for their vegetative growth, for bearing flowers and fruits and to display their full beauty. The restricted space of their root system is the main draw back of their nutrient and water reserves. Unlike it naturally grown counterparts, the roots can not spread far and wide as they like. Whenever the top soil is start to dry, watering may be given immediately to arrest the moisture stress. When watering is done, it should remember that every root of the bonsai plant should be bathed with water. All soil layers should receive their share of nutrients and water. It is best to apply water with watering-can with a fine rose can spray. The watering should be continued upto water flows out through the drainage holes present at the bottom of the pot.

During hot summers, it is necessary to give watering twice or thrice, depends upon nature of plant because of high evapo-transpiration losses. During cloudy and wet days of the monsoon, watering can be given with a few days interval. Dropping and curving of leaves indicates the moisture stress and these thirsty plants should be irrigated immediately. Heavy foliage plants may be watched carefully during hot windy days. When wilting symptoms appear these pots may be brought under the cool shade and soil should be throughly soaked. The leaves are also sprayed with cold water to decrease the temperature of aerial parts of the plant.

9.14 Time of Watering

Watering has to be done very cautiously with a power of observation. Water requirement is mainly depends upon plant species, environment, soil conditions and stage of growth. Some species like conifers, pines and cactus thrive vigourously in less water conditions. Water loving plants like *Duranta, Jamun, Bauhinia, Mulberry,* and *oranges* require frequent watering. Some plants like bongainvilleas flower profusely under slight water stress. In Autumn season, leaves will be dropped and flower buds starts to developing in peaches and plums, which require less water. In these conditions, occational watering with frequent sprinkling gives good results.

9.15 Essential tools for bonsai

- A pair of sturdy pruning clippers
- Long-handled slender shears
- Long-handled sharp scissors
- Wire cutters
- A pointed stick of cane
- Pencil shaped bamboo or wooden stick
- Copper wire
- Spatula
- Small sharp knife
- Tooth brush
- Mini compressed sprayer
- Various size & shapes of containers

9.16 Guidelines for success in Bonsai growing

- Select simple and shallow containers with sufficient drainage holes.
- Choose plants are adoptable to local weather and soil conditions
- The plant should be planted in correct position in the containers.
- Pruning and trimming should be done regularly to obtain a net work of fine twigs
- The plants should be kept in open air
- Adequate manuring should be done to ensure balanced nutrition of the plant.

- Always it is best to give mild fertilisation
- Expose the bonsai to open sun to receive plenty of solar radiation
- Air pockets should be removed when re-potting is done
- Check up the drainage holes periodically to avoid ill drained conditions.
- Aim at producing handsome or aged-looking tree.
- The plant architecture should be well-balanced and glancing naturally.
- Inspect the plants daily for nutrient deficiencies, pest and diseases.

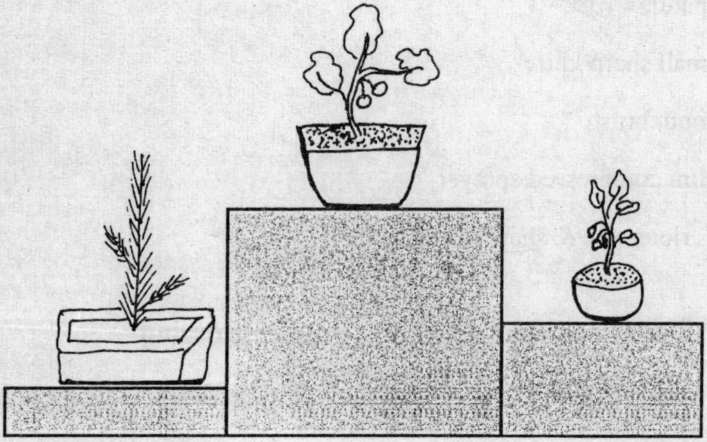

Fig 9.5: Arrangement of bonsais for displaying its beauty in the living room.

9.17 Some don'ts

- The style unsuitable to the species may be avoided.
- Plants having large leaves, flowers and fruits may be avoided to make bonsai.

- Avoid blunt cutting with implements.
- Organic manures and chemical fertilisers should not be mixed in the same solution.
- Wire used to bend the shape of the plant should not be retained for longer time.
- Bonsai should not be pampered or fattened by supplying over-rich nutrients.
- Freakish appearance of plant may be avoided.
- Do not give frequent watering to conifers
- Flowers and fruits should not be expected before the plant atains the required shape.

Epilogue

Let the graceful flowers Joyful

Let the gaint trees maintain their elegance

Let the woodlands be cheerful

Let the tiny buds, semi-opened flowers more odorous

Let the pretty leaves shine with silvered dew

Let the humming bees be more agile for collecting sweet nectare

Let the colourful birds sing loudly with melodious voice.

Let all the spirit of beautiful tree be ascended.

Let the nature smile merrily and send warm greeting in all seasons.

Appendix 1 Trees of Cool Shade for National Highways and Urban Roads

1. *Tamarindus indica* (Tamarind)
2. *ArtoCarpus heterophyullus* (Jack tree)
3. *Syzygiumn Cumini* (Jambul)
4. *Morus alba* (Mulberry)
5. *Arto carpus altilis* (Bread-fruit tree)
6. *Pongamia Pinnata* (Karanj)
7. *Prosopis cinararia* (Khejri)
8. *Quercus dilata* (Green oak)
9. *Bauhinia variegata* (Buddhist banhimia)
10. *Ficus benghalensis* (Banyan)
11. *Ficus retusa* (Ganga ravi)
12. *Ficus religiosa* (Peepal)
13. *Acaoa Sinuata* (Soap pod)
14. *Melia azadarch* (Persian Lilac)
15. *Azadirachta indica* (Neem)
16. *Madhuca indica* (Indian butter tree)
17. *Shoria robusta* (Sal)
18. *Albizia lebbek* (Sinris)
19. *Dalbergia Sissoo* (Sissoo)
20. *Artocarpus hirsutus* (Wild Jack)
21. *Artocarpus Chaplasha* (Chaplash)
22. *Kigelia pinnata* (Kigelia)
23. *Diospyros embryopteris*

Appendix 2 Trees of Graceful Foliage for the Roads of Metropolis and Towns

1. *Putranjiva roxburghii* (Putranjiva)
2. *Eucalyptus globulus* (Tasmamian blue gum)
3. *Eucalyptus gummifera* (Cask stave Eucalyptus)
4. *Eucalyptus hemiphloia* (Kino eucalyptus)
5. *Eucalyptus multiflora* (Swamp Mahogany)
6. *Eucalyptus Smithii* (Australian Eucalyptus)
7. *Eucalyptus viminalis* (Austrlian manna)
8. *Eucalyptus calophylla* (Bay kino eucalyptus)
9. *Terminalia arjuna* (Arjuna)
10. *Terminalia bellirica* (Belliric myrobalan)
11. *Terminalia alata* (Laurel)
12. *Terminalia bialrta* (Silvergrey wood)
13. *Terminalia catapa* (Indian almond)
14. *Terminalia chebula* (Chebulic murobalan)
15. *Terminalia procera* (White Bombwary tree)
16. *Terminalia paniculata* (Flowering murdah)
17. *Phyllanthus acidus* (Amla tree)
18. *Phyllanthus maderaspatensis* (Black usiri)
19. *Phyllanthus lawii* (Basket phyllanthus)
20. *Phyllanthus polyphyllus* (Tannin phyllanthus)
21. *Calophyllum inophyllum* (Alexandrian Lauril)

Appendix 3 Trees of Decorative Flowers for Highways and Approach Roads

1. *Bombax malabaricum*
2. *Cassia nodosa*
3. *Cassia grandis*
4. *Chorisia speciosa* (Mexican silkcotton)
5. *Colvillea racemosa*
6. *Lagerstroemia flos-reginae* (Queen's flower)
7. *Millingtonia bortensis*
8. *Peltophorum ferrugineum*
9. *Poinciana regia*
10. *Sterculia chlorata*
11. *Anthocelphalus indicus*
12. *Spathodea companulate* (African tulip)
13. *Delonix regia* (Gold mohur)
14. *Cassia fistula* (Indian Laburnum)
15. *Thespesia populanea* (Indian tulip)
16. *Mesua ferrea* (Ironwood tree)
17. *Mussanenda frondosa* (Srivati)
18. *Peltophorum pterocarpum* (Rusty shield bearer)
19. *Milletia ovalifolia* (Moulmein rosewood)
20. *Enterolobium saman* (Rain tree)
21. *Colvillea racemosa* (Colville's glory)
22. *Brownea Coccinea* (West Indian Mountain rose)
23. *Cassia fistula* (Indian Laburnum)
24. *Bauhinia purpurea* (Mountain abony)
25. *Java cassia* (Cassia Javanica)
26. *Cassia renigera* (Burmese pink Cassia)

27. *Poinciana regia* (Gul Mohur)
28. *Rhododendron arboreum* (Buras rhododendron)
29. *Cochlorspermum gossypium* (yellow silk cotton)
30. *Jacaranda minosaefoia* (Jacaranda)
31. *Tabebuia speciosa* (Mauve Tabebuia)
32. *Plumeria acutifolia* (Temple tree)

Appendix 4 Trees of Fragrant Flowers for Planting in Cities

1. *Dillenia indica* (Big kalinga, Eliphant apple)
2. *Dillemia pentagyna* (Small Kalinga)
3. *Dillenia andamanica* (Andaman Dillimia)
4. *Dillimia bracteata* (Small teak)
5. *Mimusops elengi* (Bullt wood tree)
6. *Mimusops littoralis* (Pinle – mohwa)
7. *Mimusops roxburghiane* (Renga)
8. *Mimusops Kanki* (Khirni)
9. *Pterospermum acerifolium* (Hathipalia)
10. *Pterospermum diverssibolium* (Mooli)
11. *Pterospermum lanceaefolium* (wild kalla)
12. *Plumeria acutifolia* (Temple tree)

Appendix 5 *Trees that yield Nectar for honey bees*

1. Purple bauhinia (*Bauhinia purpurea*)
2. Bottle brush (*Callistemon lanceolatum*)
3. Faman (*Engenia jambolana*)
4. Kachnar (*Bauhinia indica*)
5. Neem (*Azadirachta indica*)
6. Shisham (*Dalbergia sissoo*)
7. Soap nut (*Sapindus indica*)
8. Tun (*Cedrela tuna*)
9. Barna (*Cratreva religiosa*)
10. Sapium (*Sapium sebiferum*)

Appendix 6 Trees that can with stand marshy ecosystem

1. Swamp eucalyptus (*Eucalyptus rostrata*)
2. River bank willow (*Salix babylonica*)
3. Waterside willow (*Salix tetrasperma*)
4. Farash tamarix (*Tamarix* spp.)
5. Plantains (*Masa paredisica*)

Appendix 7 *Trees that can be grown in Saline and alkali soils*

1. Neem (*Azadirachta indica*) - Moderate resistant
2. Mahua (*Bassia latifolia*) - Moderate resistant
3. Dhak (*Butea frondosa*) - Extremely resistant
4. Irap Eucalyptus (*Eucalyptus citriodora*) - Moderately resistant
5. Date palm (*Phoenix dactylifera*) - Thrives in brackish soil
6. Amla (*Phyllanthus emblica*) - Slightly resistant
7. Guava (*Psidium guava*) - Slightly resistant
8. Farash tamarix (*Tamarix articulata*) - Moderate resistant
9. Bhendi (*Thespesia populnea*) - Moderate resistant

Appendix 8 Trees : that are resistant to water stress of arid regions

Name of the species	Phenology & Morphology	Degree of drought resistant
Siris (*Albizzia lebbek*)	A deciduous, spreading branches, fast-growing, 10 to 25 m. high	Moderately drought resistant
Amaltas (*Cassia fistula*)	A moderate sized, deciduous tree, 12 to 18 m. high, yellow flowering in May, tree base covered with yellow flower carpet leaves dropped before blooming	Medium drought resistant.
Palas (*Butea frondosa*)	A medium sized, deciduous tree, flowers in March after complete leaf shedding. The branches covers with scarlet flowers	Extremely drought resistant
Persian Lilac (*Melia azadarach*) or Bakain	A medium sized, deciduous, 10 to 15 m. high, Blooms in March, purple panicles appear at the end of winter	Extremely drought resistant
Malberry (*Moras indica*)	A medium-sized, deciduous, 8 to 15m. high	Moderately drought resistant
Salvadora or Mustard tree (*Salvadora persica*)	A small, ever-green tree with small, ever green with oval shaped fleshy leaves, 5 to 12m. high.	Extremely drought resistant.
Date-palm (*Phoenix dactalifera*)	A medium sized, ever-green palm, mostly found in Iran, Iraq, and Pakistan.	Extremely drought resistant.
Mesquite bean (*Prosopis juliflora*)	A medium sized, deciduous tree, gracy feather shaped foliage, rapidly growing,	Extremely drought resistant.
Beef-wood tree	A tall, ever green straight tree, 20	Extremely drought

Name of the species	Phenology & Morphology	Degree of drought resistant
(*Casuarina equisetifolia*)	to 30m. high, leaves akin to needles, thrives in dry sandy soil.	resistant.
Safeda Eucalyptus (*Eucalyptus citriodora*)	A tall, ever green, straight tree, 20 to 40m. high, exploit ground water resources.	Extremely drought resistant.

Appendix 9 *The Birds that are Perching on Flowering Trees*

Name	Scientific name	Local name	Nature of call
Black patridge	*Francolinus francolinus*	Kala teetar	Chick cheek
Grey partridge	*Francolinus pondicerianus*	Safed teetar	Kateetar
Blackbreasted Quail	*Corturmix Coromandelica*	Chanak bater	Which Which
Grey Quail	*Corturnix Corturnix*	*China bater*	*Wet-me-lips*
Jungle Bush Quail	*Perdual asiatica*	*Lowwa*	*Whi Whi*
Rock Bush Quail	*Perdual argoondah*	*Troop Lowwa*	*Whi Whi*
Grey jungle fowl	*Gallus Sonneratii*	*Jungle murghi*	*Kuck Kaya*
Red Jungle fowl	*Gallus gallus*	*Lal Jungle murghi*	*Kaya Kuck*
Common Peafowl	*Pavo Cristatus*	*Mor or Magar*	*May awe*
Saras Crane	*Grus antigone*	*Saras*	*Trumpet like*
Demoiselle	*Anthropoides Virgo*	*Karkara*	*Sonorous calls*
Common Crane	*Grus grus*	*Kulang*	*Sonorous calls*
Ringed plover	*Charadrius dubius*	*Merwa*	*Chiff – Chiff*
Goggle eyed plover,	*Burhinus oedicnemus*	*Karwanak*	*Pick – Pick*
Indian Whiskered tern	*Chlidonias hybrida*	*Tehari*	*Creek – Creek*
Common Sandgrouse	*Pterocles exustus*	*Bhat*	*Kut – ro*
Green pigeon	*Treron phoenicoptera*	*Hara Kabutar*	*Mellow Whistling*

Name	Scientific name	Local name	Nature of call
Blue Rock pigeon	*Columba livia*	*Kabutar*	*Gooter – goo – gooter*
Spotted Dove	*Streptopelia Chinesis*	*Chitroka fakhta*	*Kroo – Kruk – Kroo*
Red turtle Dove	*Streptopelia tranquebarica*	*Itwa*	*Groo – gurr – goo*
Rose ringed parakeet,	*Psittacula krameri*	*Tota*	*Kreak – Kreek*
Alexandrine parakeet,	*Psittacula eupatria*	*Bada Tita*	*Kreek – Kreek*
Koel or Kokila	*Eudynamys scolopacea*	*Koel*	*Kuoo – Kuoo*
Crow – Phesand	*Centropus sinensis*	*Jungle Kuka*	*Coop – Coop*
Great horned Owl	*Bubo bubo*	*Ghughu*	*Bu – bo*
Indian Night Jar	*Caprimulgus asiticus*	*Chhipak*	*Chuk – Chuk*
Small Blue Kingfisher	*Alcedo atthis*	*Chota kilkila*	*Chichee – Chichee*
Whitebreasted Kingfisher	*Halcyon Smyrnensis*	*Safed Kilkila*	*Chichee – Chichee*
Pied Kingfisher	*Ceryle rudis*	*Koryala Kilkia*	*Chirruk – Chirruk*
Indian Roller or Blue Jay,	*Coracias benghalensis*	*Sabjak*	*Raucous Calls*
Green Bee – eater	*Merops Orientalis*	*Patringa*	*Tit – tree – tit*
Hoopoe	*Upupa epops*	*Hudhua*	*Hoo – po – po*
Crimson breasted barbet,	*Megalaima haemacephala*	*Chota basanta*	*Tuk tuk*
Indian Pitta	*Pitta brachyura*	*Naorang*	*Wheet – tew*
Black headed oriole	*Oriolus Xanthornus*	*Peelak*	*Cheeah*
Black Drongo	*Dicruus adsimilis*	*Bujanga*	*Harsh scolding*

Appendix 10 *Flower Emblems of Various Nations*

	Flower		Country
1.	White Lily	-	Canada
2.	Beach	-	Denmark
3.	Lily	-	France
4.	Corn flower	-	Germany
5.	Rose	-	Iran
6.	Sham rock	-	Island
7.	White lily	-	Italy
8.	Chrysanthemum	-	Japan
9.	Rose	-	United Kingdom
10.	Lotus	-	India
11.	Pomegranate	-	Spain
12.	Thistle	-	Scotland

Appendix 11

Research Centers Supported by Consultative Group on International Agricultural Research (CGIAR)

CIAT - Centro International de Agricultura Tropical –Columbia

CIFOR - Center for International Forestry Research – Indonesia

CIMMYT - Centro for International de Majoramiento de Maizy Trigo, Mexico

CIP - Centro International de la Papa – Peru

ICARDA - International Center for Agricultural Research in the dry Aras – Syria

ICLARM - International Center for Living Aquatic Resources Management – Philippines.

ICRAF - International Centre for Research in Agroforestry – Kenya

ICRISAT - International Crops Research Institute for the Semi-Arid Tropics – India

IFPRI - International Food Policy Research Institute – U.S.A.

IIMI - International Irrigation Management Institute – Sri Lanka

IITA - International Institute for Tropical Agriculture – Nigeria

ILRI - International Livestock Research Institute – Nairobi

IPGRI - International Plant Genetic Resources Institute – Italy

ISNAR - International Service for National Agricultural Research – Netherlands

WARDA - West Africa Rice Development Association – Cote d' lvoire

IRRI - International Rice Research Institute, Philippines.

Appendix 12 *Cities Found on the River Banks*

	City	River
1.	Alexandria	Nile (Egypt)
2.	Amsterdam	Amsel (Holland)
3.	Baghdad	Tigris (Iraq)
4.	Bangkok	Menam (Thailand)
5.	Belgrade	Danube (Yugoslavia)
6.	Berlin	Spree (Germany)
7.	Bonn	Rhine (Germany)
8.	Budapest	Danube (Hungary)
9.	Cairo	Nile (UAR)
10.	Canton	Canton (China)
11.	Galasgow	Clyde (Scotland)
12.	Hamburg	Elbe (Germany)
13.	Karachi	Indus (Pakistan)
14.	Lahore	Ravi (Pakistan)
15.	London	Thames (England)
16.	Montreal	Ottawa (Canada)
17.	Moscow	Moskva (Russia)
18.	New York	Hudson (USA)
19.	Paris	Seine (France)
20.	Quebec	St. Lawrence (Canada)
21.	Rangoon	Irawadi (Myanmar)
22.	Rome	Tiber (Italy)

23. Tokyo	Sunida (Japan)
24. Vienna	Danube (Austria)
25. Warsaw	Vistula (Poland)
26. Washington	Potamac (USA)
27. Ahmedabad	Sabarmati (Gujarat)
28. Cuttack	Mahanadi (Orissa)
29. Gauhati	Brahmaputra (Assam)
30. Lucknow	Gomti (U. P)
31. Patna	Ganga (Bihar)
32. Srinagar	Jhelum (J & K)
33. Surat	Tapi or Tapati (Gujarat)
34. Tiruchirapalli	Cauveri (Tamilnadu)
35. Delhi	Yamuna (Delhi)
36. Hyderabad	Musi (India)
37. Vijayawada	Krishna (A.P)
38. Rajahmundry	Godavari (A.P)
39. Allahabad	Confluence of Ganga, Yamuna & Saraswati
40. Agra	Yamuna (U.P)

Appendix 13

Sobriquets

1.	Britain of South	-	New Zealand
2.	City of Skyscrapers	-	New York
3.	City of Dreaming Spires	-	Oxford (U K)
4.	City of Golden Gate	-	San Francisco (USA)
5.	City of Seven hills	-	Rome
6.	City of Magnificent Distance	-	Washington D. C
7.	Cockpit of Europe	-	Belgium
8.	Emerald Island	-	Island
9.	Empire city	-	New York
10.	Forbidden city	-	Lhasa
11.	Gate of tears	-	Bab-el-mandab, Jeruslem
12.	Great white way	-	Broad way, New York
13.	Garden of England	-	Kent
14.	Granite city	-	Aberdeen, Scotland
15.	Hermit Kingdom	-	Korea
16.	Island of pearls	-	Bahrain
17.	Land of Golden fleece	-	Australia
18.	Land of Thousand Lakes	-	Finland
19.	Land of Morning Calm	-	Korea
20.	Land of Cakes	-	Scotland
21.	Land of White elephants	-	Thailand
22.	Never Never land	-	Prairies of N. Australia
23.	Pearl of Antiles	-	Cuba
24.	Playground of Europe	-	Switzerland
25.	Quaker's city	-	Philadelphia
26.	Venice of the North	-	Stockholm, Sweden
27.	Windy city	-	Chicago
28.	Worlds bread basket	-	Prairies of N. America

Appendix 14

Romman Numerals

I	=	1
II	=	2
III	=	3
IV	=	4
V	=	5
VI	=	6
VII	=	7
VIII	=	8
IX	=	9
X	=	10
XI	=	11
XII	=	12
XIII	=	13
XIV	=	14
XV	=	15
XVI	=	16
XVII	=	17
XVIII	=	18
XIX		19
XX	=	20
XXX	=	30
XL	=	40
L	=	50
LX	=	60
LXX	=	70
LXXX	=	80
XC	=	90
C	=	100
CC	=	200
CCC	=	300
CD	=	400
D	=	500
DC	=	600
DCC	=	700
DCCC	=	800
CM	=	900
M	=	1000
MM	=	2000
MMI	=	2001

Appendix 15

Metric System of Measures

a) Linear

1 Millimetre	=	0.039 inch		
1 Centimetre	=	10 mm	=	0.394 inch
1 Decimetre	=	10 mm	=	3.94 inch
1 Metre	=	100 cm	=	1.094 yards
1 Decametre	=	10 m	=	10.94 yards
1 Hectometre	=	100 m	=	109.4 yards
1 Kilometre	=	1000 m	=	0.6214 mile

b) Capacity

1 Millilitre	=	0.002 pint		
1 Centilitre	=	10 ml	=	0.018 pint
1 Decilitre	=	10 cl	=	0.176 pint
1 Litre	=	100 cl	=	1.76 pints
1 decalitre	=	10 litres	=	2.20 gallons

c) Weight

1 Milligram	=	0.015 gram		
1 Centigram	=	10 mg	=	0.154 gram7
1 Decigram	=	10 cg	=	1.543 gram
1 Gram	=	10 dg	=	15.43 grams
1 Decagram	=	10 g	=	5.64 drams
1 Hectogram	=	100 g	=	3.527 ounces
1 Kilogram	=	1000 g	=	2.205 pounds
1 Tonne	=	1000 kg	=	0.964 long ton

d) Cubic measure

1 Cubic Centimetre	=	0.061 Cu. inch		
1 Cubic metre	=	1.306 Cu. inch		

e) Square measure

1 Square Centimetre	=	0.155 Sq. inch		
1 Square metre	=	1.196 Sq. yards		
1 are	=	100 Sq. metres	=	119.6 Sq. yards
1 hectare = 100 ares	=	2.471 acres		
1 Sq. Kilometre	=	0.386 Sq. mile		

f) Temperatures

Celsius	Fahrenheit
- 17.8°	0°
- 10°	14°
0°	32°
10°	50°
50°	122°
90°	194°
100°	212°

Glossary of Terms

ACCRESCENT	: The floral parts of calyx or corolla that persist and enlarge after flowering
ACHENE	: A small dry indehiscent, single seeded fruit.
ACICULAR	: Needle shaped
ACORN	: The nut enclosed within a cup formed by thickened bracts.
ACULEATE	: Covered with prickles
ACUMINATE	: Terminating with a sharp tapering point
ACUTE	: Evenly tapering without any prolongation
ADPRESSED	: Coming close against the producing surfaces
AESTIVATION	: The mode of floral parts folded in the bud and their relative position.
AMPLEXICAUL	: The base of petiole clasping the shoot
ANDROGYNOUS	: Inflorescence bearing both male and female flowers.
ANNULATE	: A mark of rings appear transversely.
ANTERIOR	: Floral part inserted away from the axis of inflorescence
ANTHER	: Tip of stamen that bears pollen
APICULATE	: Having short pointed tip.
AMPLEXICAUL	: The base of petiole when clashing the stem.
APOCARPOUS	: Carpels are free and distinct
ARBORESCENT	: Growing to a size of tree.
ARCUATE	: Moderately curved like a bow

ARIL	:	An accessory covering the seed
ARISTA	:	Bearing a bristle or own
ASCENDING	:	Upward direction
AURICULATE	:	Consists of flat lateral expansion
AXIL	:	The upper angle formed between leaf and supporting stem.
BACCATE	:	The upper angle formed between leaf and supporting stem.
BACCATE	:	The fruit berry having pulp like substance
BAST	:	Fibrous inner bark of the stem.
BERRY	:	A succulent fruit with no stone but having more than one seed.
BIFARIOUS	:	In two opposite verticle rows
BIPINNATE	:	Twice pinnate
BLADE	:	Lamina or flat part of leaf
BULBIL	:	A small bulb produced in the leaf axil or on the inflorescence
BIFURCATE	:	Having two prongs or branches
BILABIATE	:	Having two lips
BISERIATE	:	Two rows are formed within the other.
BOLE	:	The main axis of a tree
BRACT	:	A rudimentary leaf subtending a flower
BULB	:	A conical stem, flattened at the base and conical above which bears buds.
BUTTRESSED	:	Stems having vertical ridges or projections.
CADUCOUS	:	Falling off very early
CAESPITOSE	:	Growing in tufts like bamboos.
CALLUS	:	Tissues formed on the wounds of plants.
CALVOTRA	:	A hood like cap
CAMPANULATE	:	Shaped like a bell

CANESCENT	:	Pubeseence produce a greyish white hue on epidermis.
CAPITULUM	:	A dense head of flowers.
CAPSULE	:	A dry syncarpous fruit
CARTILAGINOUS	:	Tough and hard like parchment
CARUNCLE	:	Excrescence at the micropyle and of some seeds like castor.
CARYOPSIS	:	Seed like fruit of poaceae family.
CAUDATE	:	Having a slender tail like terminal appendage.
CHARTACEOUS	:	The thin and flexible texture like paper.
CILIATE	:	Fringed like the hairs of eye-lashes
CIRRHOSE	:	Having tendrills
CLADODE	:	A flattened branch assumes the function of a leaf
CLAVATE	:	Thin below and thickened upward gradually like club.
COLLETERAL	:	Standing side by side.
CORDATE	:	Like the figure of heart on cards.
COME	:	A tuft of soft hairs on a seed
COMOSE	:	Having a coma.
CONCOLOR	:	Of the same colour throughout
CONDUPLICATE	:	Folded once upon itself length wise.
CONNATE	:	Parts of the same organ when congemitally united
CONE	:	The multiple fruit of combers
CORIACEOUS	:	Tough and thick like leathery texture.
CORYMB	:	The pedicels of flowers borne from various points but attain the same level on top.
CRENATE	:	Having round teeth.
CRUSTACEOUS	:	Covering with hard and brittle substance.
CULM	:	A hollow and Jointed stem of grasses.
CUNEATE	:	Sharp angles at the base similar to wedge.

CYME	:	A dry inferior achene with adnate calyx which is found in the fruits of *Asteraceae*.
DECIDUOUS	:	Leafless for a part of the year.
DECOMPOUND	:	Divided more than once
DEHISCENCE	:	A mode of opening of a capsule
DENTATE	:	Minutely teethed
DEXTRORSE	:	Twining from left to right.
DIADELPHOUS	:	The stamens are united by their filaments into two bundles.
DIFFUSE	:	Spreading the branches loosely
DIGITATE	:	A finger like compound leaf with a common petiole
DIOECIOUS	:	The male and female flowers are present in different plants.
DIVARICATE	:	Spreading widely apart
DRUPE	:	A stony fruit with pulpy pericarp and bony endocarp.
ECHINATE	:	Covered with sharp bristles like a hedge hog.
ELLIPTIC	:	Oblong or oval with rounded ends.
ENSIFORM	:	A shape of sword
EXCAVATE	:	Hollowed into broad deep pits
FALCATE	:	Curved like a blade of schythe
FILIFORM	:	Thread like structure
FISTILAR	:	Hollow
FLABELLATE	:	Fan-shaped
FLEXUOUS	:	Wavy shaped, Zig- Zag.
FARINACEOUS	:	Consisting of resembling starch
FAMILY	:	A group of Co-related genera or tribes
FASCICLED	:	In dense clusters
FERRUGINOUS	:	Coloured like iron rust
FIBROVASCULAR	:	A tissue consisting of woody fibres and ducts.

FLOCCOSE	:	Wooly with locks of dense soft hairs.
FUGACEOUS	:	Falling off or fading early.
FURCATE	:	Forked
FUSIFORM	:	Spindle shaped and tapering at both ends.
GIBBOUS	:	Having a pouch like swelling.
GLAUCOUS	:	A bluish-grey colour covered with fine bloom.
HASTATE	:	Shaped like spear
HERMAPHRODITE	:	Both stamens and pistil are present in the same flower.
HIRSUTE	:	Covering with long and coarse hairs.
HYALINE	:	transluscent.
IMBRICATE	:	Over lapping like tiles on the roof of house.
IMPARIPINNATE	:	Unequally pinnate.
INDEHISCENT	:	Opening in a irregular manner.
INFUNDIBULIFORM	:	Funnel shaped
INVOLUTE	:	Rolled inward
LANCEOLATE	:	Shaped like lance head and narrowed at both ends.
LAX	:	Flowers are not closely arranged or diffuse in the inflorescence.
LENTICULAR	:	Double-convex lense shaped
MARCESCENT	:	Withering of part without falling on the plant.
MONOECIOUS	:	Male and female flowers are present on the same plant
MURICATE	:	Rough surface with hard tubercles
OBLANCEOLATE	:	Long narrow and tapering at both ends but broader in the middle.
ORDER	:	A group of closely related families.
PAPILIONACEOUS	:	Butterfly like flowers consists of keel, wings and standard petals.
PAPPUS	:	Achenes and fruits consists of hairy tufts.

PATENT : Widely spreading.

PARIPINNATE : Pinnate without terminal leaflet.

PEDICAL : Ultimate stalk supporting a single flower.

PENULTIMATE : Next to ultimate.

PEDATE : Like a birds foot, palmately divided with lateral divisions 2-deft.

PINNATE : Leaflets arranged on either side of axis like webs of a feather on its shaft.

POME : A succulent fruit having seeds embedded in many cartilaginous cells as in the pear.

PAPILOSE : Bearing minute nipple – shaped projection.

PENDUNCLE : The stalk supporting a solitary or cluster of flowers or fruits.

PELLUCID : Transparent or translucent.

PERIANTH : A floral envelope when calyx and corolla are similar form.

PERSISTENT : Remaining attached until the fruit is mature.

PROCUMBENT : Lying flat on the ground

PUBERULOUS : Sparcely pubescent

PYRIFORM : Pear-shaped

RADICLE : Axis of an embryo below the cotyladons.

RADICAL : Proceeding from the root.

RENIFORM : Kidney shaped

RUGOSE : Full of wrinkles

SAGITTATE : Shaped like a arrow head

SARMENTOSE : Producing long straggling woody branches.

SCABROUS : Rough to touch due to presence of stiff hairs.

SESSILE : Attached directly without a stalk.

SIGMOID : Curved like greek letter sigmas.

STIPULE : An appendage of leaf on either side of petiole

STRIGOSE : Covered with sharp pointed hairs.

TRICHOTOMOUS	:	Dividing into three always.
UNDULATE	:	Wavy margin.
VERTICILLATE	:	arranged in whorls
VIVIPAROUS	:	The seeds germinate before falling on the ground.
VALVATE	:	With the margins of members of a whorl only meeting without overlapping.
VARIETY	:	A group of individuals differing from others of the same species.
VENATION	:	The mode in which veins are disposed in a leaf.
VENTRICLE	:	A lateral swelling or bulging out
VENTRICOSE	:	Having a ventricle and unequally swollen.
VERRUCOSE	:	Covered with wart like excrescenes.
VERSATILE	:	Swinging to and fro.
VILLOUS	:	More or less thickly covered with long, soft hairs.
VIRGATE	:	Twiggy or rod like.
VISCID	:	Sticky.
WING	:	A thin membranous appendage.
WHORL	:	Similar members that are arranged in a circle round an axis.
XYLEM	:	A woody tissue formed by the inner portion of fibrovascular bundle
ZYGOMORPHIC	:	Flowers which can be divided into similar halves in one plane only.

Selected References

1. Abraham, Z (1981) Ethanobatany of the Todas, the Kotas and the Irulas of the Nilgiris. *Economic Botany* 32: 310-312.

2. Agarwal, V.S. (1986) *Economic plants of India*, Kailash Prakashan, Calcutta.

3. Anonymous (1962). *Wealth of India*, publications and information Directorate, Govt. of India, Vol VI: pp 208-215.

4. Anonymous (1988). *Wealth of India – Raw materials* Vol 1-11, CSIR, Pusa, New Delhi.

5. Anonymous (1991) *Bamboo – The poorman's Timber*, Indian Council of Forestry Research & Education, New Forest, Dehradun. 24P.

6. Bhargava, N (1981). *Plants in the Folk life and folklore in Andamans & Nicobar Islands* (In: Glimpses of Indian Ethanobotany. Ed. S.K. Jain).

7. Dhanda Leila (1984) *Bonsai culture* Oxford & IBH publishing house, New Delhi., 179 P

8. Dwivedi, A.P (1990) *Silviculture – Principles and practices*, Surya publishers, Dehra Dun, 458

9. Jain, A.K (1989) *Forest Economy of India*, Vohra Publishers & Distributors, Allahabad, 283 P.

10. Ovington J. D. (1965) *Wood lands*. The English University press LTD, London, 286 P

11. Raju, R.A. (1997) Forest Wealth of India – inaccessible to impoverished. Daya Publishing house, Delhi. 320 P

12. Raju, R. A 1997. *Revitalizing health Through vegetables*. Agrotech Publishing Academy, Udaipur 313002, 128 P

13. Raju, R. A (2000) *Ethnobotany of Rice weeds in South Asia*. Today & Tomorrow's Printers & Publishers New Delhi. 235 P

14. Raju, R.A. (2000) *Wild plants of Indian Subcontinent* and *Their economic use*. CBS publishers & Distributer, New Delhi 110002 200 P

15. Randhawa, M. S (1996) *Flowering Trees*, National Book Trust, New Delhi, 208 P

16. Santapau, H (1993) *Common Trees*, National Book Trust, New Delhi, 118 P

17. Sagreiva, K. P (1994) *Forests and Forestry*: National Book Trust, India; New Delhi, 319 P

18. Singh, R.V (1988). *Forestry and food security in India*, FRI, Dehradun, 118 P

19. Singh, U, wadhwani, A M and Johri, B M (1985) *Dictionary of Economic plants in India*, ICAR, New Delhi.

20. Solanki, M S 1981. *Forests as a source of food*. Vol I and II, FAO Regional office, Bangkok.

21. Troup, R. S 1975. *The Silviculture of Indian Trees* (Revised) FRi, Dehradun, 307 P.